TAKING THE COUNTRY'S SIDE

AGRICULTURE
AND ARCHITECTURE

SÉBASTIEN MAROT

Lisbon ArchiTecTure Triennale
The PoeTics of Reason

CROSSED TERRITORIES

For centuries the urban and rural worlds have lived side by side and only in recent decades have they gradually been forced into an extreme dichotomy. This phenomenon, an outcome of the mass mass migration into our cities, has led to the gradual abandonment of the rural world and to an imbalance in our ecosystem that once integrated nature and human settlements and activities. Urban centres have spread exponentially in a fragmented way, and population figures have grown exponentially as life expectancy rates have improved. Meanwhile rural environments, which once thrived, have been left to suffer neglect or the installation of giant agri-industrial corporations, the latter often with brutal consequences for the environment.

What we are dealing with is a perfect storm that is the combined result of exacerbated and uncontrolled consumerism, a lack of ethics on the part of producers and distributors of consumer goods, and the negative effects of fuel and energy corporations, whose products are extremely harmful in terms of production and consumption. In many countries this is all happening with the consent, or at least disregard, of the regulatory agencies, which are supposed to tame the markets, and governments, whose job it is to set up and monitor the work of said agencies and the respective markets.

Meanwhile, communication systems have become highly effective, both in terms of negotiation channels and the transport of goods, with the globalisation of markets on a worldwide scale, as opposed to the ancestral model based on proximity. The global economy is now based on out-of-control consumerism, whether it is superfluous goods such as food stuffs, which are now available out of season; or clothing, with products becoming rapidly and intentionally out of fashion in swiftly changing seasons.

A central question is posed by the current paradigm of voracious consumption of environmentally unfriendly fuels and energy that comes from the need to cover ever-larger distribution distances and serve ever-growing urban areas. If we are to save the planet, much more than urban and regional planning models will have to change. An urgent need for a change in paradigm through a return to the principle of urban density has been identified. In addition to promoting economic and social sustainability, this would also shorten product distribution distances and enable the more widespread adoption of cleaner, renewable energy sources. However, this is only a partial solution that does not take all aspects of the issue into consideration.

Agriculture and Architecture: Taking the Country's Side, a reflection based on the exhibition organised in Garagem Sul by the Lisbon Architecture Triennale 2019 under its theme The Poetics of Reason, explores the possibilities created by various urban and regional planning models on the basis of the rich history of environmentalist movements.

```
José Mateus
Executive Chairman
Lisbon Architecture Triennale
```

Translated by Liam Burke.

SÉBASTIEN MAROT

INTRODUCTION

"LET NO ONE ENTER HERE WHO IS IGNORANT OF GEOMETRY"

Such was the warning writ large on the gate of Plato's Academy, summoning *logos* (reason) and sound dialectics in the discussion of major topics and public affairs. A qualified version of this sentence should also be placed at the entrance of this exhibition: Let no one enter here who is, and intends to remain, ignorant of both the scale and limitations of the biosphere. Indeed, the said ignorance is so pervasive nowadays, especially in the spheres of mainstream politics and planning, and so apt to be paraded as commonsense, that it deserves to be treated at least with some degree of scorn if not completely ostracised.

SCHIZOPHRENIA

Any reflexive and reasonably informed person today, who takes time to ponder on the global dynamics of our world, is confronted with a highly perplexing situation. When looking back at the past, it seems as if the urbanisation of our planet, which has steadily been increasing over the past two or three centuries, is inevitable and is integral to history. Indeed, scores of official statistics demonstrate that the global population will keep growing, at least in the present century, and will most likely concentrate, as it does now, in larger and larger metropolises. On the other hand, when one probes the future, and the environmental issues that loom there, such as climate change, fresh water scarcity, soil erosion, peak oil and biodiversity collapse, this same urbanisation looks impossible.

Such a paradoxical situation (both inevitable *and* impossible) is extremely conducive to schizophrenia, and confronts reason with an almost unbearable challenge. So unbearable in fact, that people generally choose, consciously or not, to overlook or ignore one side of the equation. Taking up Charles Mann's categories, one might divide them into two classes: the wizards and the prophets[1]. For wizards, the hard facts are statistical tendencies; representative of the way that people have tended to behave and will most probably continue to do so. Hence, the task of science, technology and politics is to manipulate and trick reality so as to clear and smooth the way for the inevitable. In their view, history has always been made possible by those who fought to overcome, in the short- or medium-term, the predicament of mankind, and postpone the obvious limits to its development.

Rem Koolhaas's idea of the metropolis as an accumulation of potential catastrophes that never happen, and the plea for superurbanism which he put forward in *Delirious New York* as a way of overcoming "reality shortage", stands out as a masterful example of the philosophy of wizards. Prophets, on the other hand, tend to focus on the limits that others are eager to push and overcome. They typically stress that the carrying capacity of our planet indeed has limits, however difficult to ascertain precisely, and that trying to trick them in the short or mid-term can only lead to their asserting themselves more violently in the end. In their view, human societies should urgently learn to live and prosper within these limits and the natural metabolism of our biosphere.

Whereas wizards think that the task of reason (namely science and technology) is to outsmart and disrupt those apparent limits, prophets claim that its role is to wisely acknowledge them and learn to live in good intelligence with them. Unlike Charles Mann, who refrains from taking sides between these two positions, presenting them as irreconcilable philosophies, bound to constantly oppose and correct each other, our ambition here is clearly to present the case for prophets and explore the implications of their concerns and ideas for both agriculture, architecture and urbanism.

---

[1] **Charles C. Mann, *The Wizard and the Prophet: Science and the Future of Our Planet*, Alfred A. Knopf, 2018. The two main figures that Mann brilliantly contrasts in this book are William Vogt (the prophet), who famously launched the wave of postwar environmentalism with his groundbreaking *The Road to Survival* (1948), and Norman Borlaug (the wizard) who was awarded the Nobel Peace Prize in 1970 for his key role in the Green Revolution. Significantly, both developed their arguments from a deep consideration of the state and predicaments of agriculture.**

NEXUS
The basic hypothesis of this book and exhibition is that no sound reasoning will develop on the future of agriculture and architecture, which both emerged as the twin fairies of the Neolithic revolution (and thereafter of the Anthropocene), unless those two fields of concerns, and their associated modes of living, are reconnected and fundamentally rethought in conjunction with one another. Our intention is thus to deeply question the growing divorce and estrangement of the two disciplines, as it was initiated by the scientific revolution (and its so-called mastery and domination of nature), pronounced by the spread of the market economy, and consecrated by the industrial era, which launched both into the parallel dead-ends of metropolitan congestion and monocultural deserts.

Taking the Country's Side extends to architects, as well as to all those concerned by the current evolution of our living environments, an invitation to leave their metropolitan niche, their zones of professional comfort and smartness, and literally take a walk on the wild side. For several decades now, several individuals and communities, committed to enacting alternatives to the deleterious processes of industrial agriculture and market economy (under the name of permaculture, social ecology, agroforestry, bioregionalism or agroecology), have evolved a treasure trove of ideas and principles that significantly challenge the core concepts of architecture and urbanism today. As a poetics of reason for the Anthropocene, this practical wisdom is in our view much more pointed and precious than what academia generally has to offer on these issues.

NAVE
The structure of our exhibition *Agriculture and Architecture: Taking the Country's Side* was suggested by the plan of Garagem Sul which roughly mirrors, with its two long rows of pillars, that of a basilica or cathedral: a nave flanked by two aisles. This plan led us to organise our argument in three parallel components.

Our core argument, displayed in the nave, is didactic and consists of six rows of vignettes (7 in each) presenting significant ideas, moments and figures which any visitor or reader might bear in mind when considering the nexus of agriculture and architecture. Taken together, these 42 discrete panels, chronologically arranged along six thematic lines of thinking, are like a giant deck of cards in which the visitor's mind is invited to wander as in a game of patience and reflection. If chronology rules the location of these vignettes within each section, it also determines the progression from one section to the next, as the visitor, thus equipped with a rear-view mirror, moves along the nave towards the choir or apex, and focuses on our present condition.

The beauty of cards is that they have a back, whose function is generally to be mute and to conceal their true value. Here, we have used those backs to call up, as if from the womb of architecture's collective imagination, a number of famous projects, images and references that speak by themselves and need less explanation. No fixed rule dictated their placement, other than their counterpoint or resonance with the front panels situated either behind or across from them. In the darkness of Garagem Sul, they will thus float as a silent constellation of well-known references, ready to take up new meanings and connotations. In the book, these rear images are systematically associated with their corresponding fronts so most of the associations we arranged in the exhibition are blurred or lost, as happens when one collects back the cards from playing patience into the deck. But this, if anything, should only invite the reader to shuffle the deck, start reflecting, and (dis)play her own game.

CHOIR
History and theory are useless and futile if they are not aimed at understanding where we stand today and at finding ways into the difficult arena of the present. Our intention was therefore to equip our visitors and readers with a jurisprudence that might help them consciously play their game on this stage, and in the gestation of "our common future". Our 42 panels — both front and back — could thus be seen as an elaborate Way of the Cross leading up to the seventh, ultimate section that constitutes the choir of our exhibition. There, with illustrator Martin Étienne, we have tried to synthesise in four wide landscape drawings, the different and competing directions that the dialectic of city and country as well as agriculture and architecture, might take nowadays and in the near future.

Incorporation, Negotiation, Infiltration, and Secession are framed as the cardinal points of a wind rose within which visitors and readers are kindly encouraged to reflect on what their own position might be. If all these different narratives are not necessarily incompatible with one another, if they all have a basis in fact, and may present themselves as ways of confronting the present environmental predicament, one should understand however that they strongly differ on the nature of this predicament, and that their confrontation indeed defines an arena of dissent and conflict. Our intention was to make this confrontation explicit, and to stress polar opposition in Incorporation and Secession as a fierce struggle about the very concept of rationality today.

Certainly — even though we refrained from caricaturing any of our narratives — our own position is far from neutral in this arena. In the climate of collective obnubilation that has been reigning for decades, in which all our mental needles have been fixed upon cities and the metropolitan condition as upon some kind of absolute North, the hypothesis of secession (away from the ethos of urbanism and large scale land-use planning) is in our view the only one to display an adequate measure of historical imagination, i.e. the only one to take stock of the *adventure* that the present environmental predicament represents, and to lead from that by consequences. Informed by the jurisprudence shown in the exhibition and this book, our position within the magnetic field could be thus defined: right next to Secession, leaning toward Infiltration, with a modest and conditional tolerance for Negotiation, and an instinctive distrust of Incorporation.

AISLES

Even though they could not be included in this book, it may be worth saying a word about the two other components of the exhibition, displayed in the aisles of Garagem Sul. On the left-hand side, where the said aisle is like a giant corridor, we took advantage of its uninterrupted wall to display a 60 meters fresco; a timeline synthesising the parallel evolutions of agriculture and architecture (and subsequently urbanism) since their common inception in the Neolithic age. In this work-in-progress, populated with quotes from anthropologists, historians and others, we strived to highlight and relate significant environmental, technological and socio-political turning points, which changed the ways in which human societies shaped and transformed their surroundings, all the way through the industrial era, and up to the highly perplexing situation which is now ours. We thought it necessary to provide our visitors with this historical background, however imperfect, for two reasons: firstly to help them locate the different vignettes and moments of our thematic jurisprudence in a common global history, and secondly to draw their attention toward the different dynamics and temporalities of those parallel evolutions (and their interplay), which is indispensable if we are to dispel the cloud of false promises that surrounds us today and seriously address the ultimate Kantian question: What are we reasonably entitled to hope?[2]

On the right-hand side, the aisle of Garagem Sul is divided into several cells or chapels which we have used as a kind of movie multiplex, displaying reels of film excerpts, documentaries and interviews, all directly related to the topic and subthemes of our exhibition: the nexus of agriculture, architecture and urbanism; the transformations and industrialisation of farming; the different aspects of the environmental predicament and the crisis of cities; and the local experiments that have been undertaken, here and there, by committed individuals in community building, agroecology or permaculture. If collecting a sound historical jurisprudence is essential to understanding our situation today, nothing is more invigorating than a conversation with those who have indeed taken the country's side and devoted their lives to managing worlds.

MUNDUS

"World=City": such was the formula by which Rem Koolhaas, in 2000, summarised the idea of Mutations, an exhibition in which he documented the massive effects of globalisation and the accelerating urbanisation of our planet[3]. Indeed, as the term clearly indicates, what globalisation achieved is to turn the term "world" into an exclusive singular: *The* World. Fueled concomitantly by the expansion of the market economy and the massive use of dense fossil fuel energy (indispensable to the wide circulation of goods and people),

---

2   This illustrated frieze or timeline, a work in progress, could be seen as an homage to Lewis Mumford (1895–1990). Nothing is more urgent in our view than to synthesize, update, expand and refine the ground-breaking work that he undertook with *Technics and Civilization* (1934) and *The Culture of Cities* (1938) and relentlessly carried on until his death.

3   As we write these lines, we are of course well aware that Koolhaas, having shifted his gaze on the countryside and on the tremendous evolutions which are currently happening in rural areas worldwide, is about to critically revisit this whole question. Indeed, *Taking the Country's Side* may be seen, and has notably been conceived, as *an extensive footnote and contrapunctus* to the exhibition he will soon present at the Guggenheim Museum in New York.

globalisation is the autocatalytic process by which places and territories increasingly invested in their "comparative advantages" and specialised in specific skills, services, resources or amenities, thus becoming more and more dependent of all others for the rest, i.e. more and more worldless by themselves, if not frankly unworldly. In this situation, one has a world or is a "citizen of the world" in the extent to which one has secured the means and power to navigate à la carte between those highly specialised enclaves. While many Postmodern intellectuals complacently endorsed this process by singing the alleged virtues of mobility and nomadism, urbanists and planners treated cities and territories as a vast plumbing of smart grids and multimodal interconnections, an infrastructure of fluxes and disjunction. This is an insult to all those who, lacking the means or desire to use it, are actually trapped into the margins and leftover spaces of this expanding network. In the condition of environmental turmoil which is now ours, where everything indicates that the massive resources that fueled globalisation might soon dwindle, questioning the current disassociation of agriculture and architecture, cityside and countryside, basically amounts to raising the following political question: What is *a* world?

As for us, the best response we have managed to come up with so far goes approximately like this: a world is an area, territory or region, a *country*, both physical and cultural, *where one could reasonably imagine to spend and project the entirety of one's own existence*, because its disposition and configuration and the modes of coexistence that have evolved there have turned it into a living environment, sufficiently varied and complete in its genre.

Implied here is the idea that a world is a relatively autonomous region that could eventually sustain itself, i.e. a viable territory, able to satisfy a minimum, and hopefully even more, all the "basic functions" that the industrial era progressively separated, and that modern planning ultimately segregated. Obviously, at least in most latitudes, very few areas would even faintly appear to satisfy that demanding definition. This, given the very likely imminence of both energy descent and biological erosion, should be a matter of great concern and worries. But more than that, it should be an incredible incentive for young architects and designers to effectively take the country's side, *become natives*, and learn from those who have taken upon themselves to actually engage, build up and manage viable islands of coexistence and resilience. This book and exhibition have no other goal than to back up their impatience with an illusory status-quo and to help them embrace this issue with their feets, hands, minds and friends. Incidentally, we hope they will also clarify what was maybe still elliptical in the conclusion we once gave to a little essay on *sub-urbanism*: "Ours is no longer the century of expanding cities, but rather of deepening territories. No more than the simulacra of literal memory, modern nomadism will never make bearable the flattening-out of places, their increasing univocity. The earth has become too small for us even to dream of not exploring, everywhere, its fourth dimension. It is urgent to extrapolate."[4]

Sébastien Marot
July 2019

[4] Sébastien Marot, *Sub-Urbanism and the Art of Memory,* AA Publications, London 2003. In two recent essays, we have also tried to relate our reflections on the nexus of architecture and agriculture (and on permaculture) with this plea for sub-urbanism: "De l'Art de la mémoire à l'art d'espérer", in P. Mantziaas et P. Vigano, *Urbanisme de l'espoir: Projeter des horizons d'attente,* Métis Presses, Genève 2018, and "L'Envers du décor" in Augustin Rosenstiehl (ed.), *Capital Agricole: chantierspour une ville cultivée,* Pavillon de l'Arsenal 2018.

A.

AGRI CULTURE & ARCHI TECTURE

In which the bemused reader discovers that these two practices have evolved in parallel since their common cradle in the Neolithic period; that there were strong links and symbolic correspondences between them (which were progressively overlooked, repressed and forgotten); and that their reconnection may be one of our most urgent tasks.

1, 2
Plan and view of Neolithic Çatal Höyük, Anatolia.
James Mellaert, 1975

3
Neolithic village of Abu Hureyra, Northern Syria, circa 7200 BC.
Andrew Moore, 2000

HOMO DOMESTICUS

THE TWIN SISTERS OF THE NEOLITHIC DOMUS COMPLEX

The geneses of both architecture and agriculture may be traced back to the so-called neolithic revolution. The said revolution was initially defined by Australian archaeologist Vere Gordon Childe (1892–1957) as the period when groups of humans started to cultivate plants and breed animals; a profound change that would have led groups of hunter-gatherers to progressively settle and become sedentary, thus giving birth to solid buildings and villages. In fact new archaeological finds and theories have since questioned this idea by showing that permanent settlements preceded the advent of agricultural practices, especially in the so-called Fertile Crescent (Levant, Syria, Turkey and Iraq) still considered the most ancient cradle of the neolithic transition.

In other words, if some have considered agriculture to be the driving force of the neolithic transformation, in which societies shifted to higher sedentism, others insist that the real factor was the ideology of home and house making (i.e. the emergence of an elaborate architecture).

Regardless of what came first, one can note that architecture (the construction of permanent buildings and villages) and agriculture (the production of food and fibers through the explicit selection and husbandry of plants and animals) appeared as two complementary phases of an autocatalytic process of coevolution, through which larger and larger human communities increasingly designed and modified their surroundings, and coalesced into what James C. Scott has called "multispecies resettlement camps".

The Neolithic was a transition towards *domestication*, where the taming of plants and animals by humans went hand in hand with the domestication of humans themselves within permanent buildings and intensely organised territories. Indeed, if permanent or semi-permanent settlements preceded agriculture, their construction was always more or less circular and half dug in the ground (similar in plan to that of nomadic camps). What the neolithic revolution first produced in Syria shortly before 8000 BC was *rectangular architecture*. Henceforward, from this world premiere of rectilinear structures, humans and their resources would generally be housed in square modular structures, more conducive than rounded shapes to accretion, expansion and interdependency.

References
Peter Bellwood, *First Farmers: The Origins of Agricultural Societies*, Blackwell, 2004.
Jacques Cauvin, *Naissance des divinités, naissance de l'agriculture*, Flammarion, 1994.
James C. Scott, *Against the Grain: A Deep History of the Earliest States*, Yale, 2017.

4
Neolithic settlement of Beidha, Jordan.
Phot. Andrzej Dadak, 2016

5
Neolithic rectangular walls, Aşıklı Höyük, Cappadocia.
Phot. Katpatuka, 2009

PRIMITIVE HUT?
Essai sur l'Architecture, Marc-Antoine Laugier,
engraving by Charles Dominique Joseph Eisen, 1755.

1
Granaries, Combarro, Galicia.
Rudofsky, 1964

2
Ventilation grids:
Granaries, Soajo, Galicia.
Rudofsky, 1964

3
Parthenon's Doric frieze, Athens.
Phot. Hermer, 1916

AGRICULTURE AND ARCHITECTURE

SUBLIMATION

TEMPLE / GRANARY

A powerful symbol of the Athenian democracy, the doric temple (e.g. Pericles's Parthenon) stands as an embodiment of architecture, at least in the West. But where does this type come from, and what did it exactly *represent*? The classical interpretation (Vitruvius, Laugier, Choisy) holds that the temple was a mineral transposition of a public building that would have previously been made in wood. The columns stood for stilts and the frieze was a stylised wooden floor structure with triglyphs representing the end of beams. Obvious, no?

Yet in the 1980s, German architectural historian Goerd Peschken (b. 1931) challenged this interpretation. Following an intuition first described by architects such as Hans Soeder (1891–1962), Peschken developed another — much more fertile and convincing — hypothesis. Showing numerous examples of traditional vernacular granaries across Europe, some still in use the 1960s (such as those in Galicia which Bernard Rudofsky had documented in his famous *Architecture without Architects* book and exhibition from 1964), he suggested that the doric temple was basically a monumental transposition of those granaries where the treasure of the community — its stock of grains and seeds — was kept and preserved from one year to the next.

From this perspective, while the columns stand for the pilasters or stilts that protected the grain from humidity, and their capitals for the usual defenses against pests, the frieze does not represent a floor structure but the stylised compression of the granary itself, i.e. a whole story regularly pierced by ventilation grids; hence the triglyphs.

Of course, the temple was *not* a granary. But that is the whole point. As the Greek *polis* became more and more committed to, and dependent on the, imports, conflicts and enslavements that came with being a more international, maritime economy, it invented the peripter to *sublimate* (i.e. evacuate) this reality in a monumental celebration of the autonomy, foresight and independence of village communities.

SÉBASTIEN MAROT

An early — and troubling — example of *incorporation*.

References
Bernard Rudofsky, *Architecture without Architects*, Doubleday, 1964.
Hans Soeder, *Urformen der abendländischen Baukunst in Italien und dem Alpenraum*, DuMont Schauberg, 1964.
Goerd Peschken, *Demokratie und Tempel: die Bedeutung der dorischen Architektur*, Verlaag der Beeken, 1990.

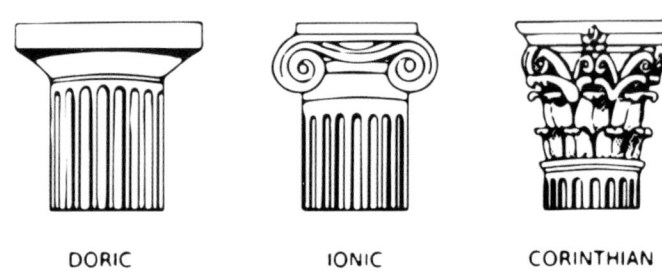

4
Triglyphs:
Temple of Poseidon, Paestum.
Peschken, 1990

5
**Classical orders and their capitals:
Three pest-control techniques?**

Silo à grain.

TROIS RAPPELS
A MESSIEURS LES ARCHITECTES

I

LE VOLUME

MODERN ARCHITECTURE AND GRAIN ELEVATORS
Vers une architecture, Le Corbusier, 1923.

1
Engraving of Marcus Vitruvius
Pollionis (circa 80–15 BC).

2
Engraving of Marcus Terentius
Varro (116–27 BC).

COINCIDENCE

A LESSON IN BOOK BINDING

The most influential treatise on architecture in the Western tradition, and a major source of information on Graeco-Roman science and technology, including the fields of engineering, infrastructure and surveying, was Vitruvius's *De Architectura* (Ten Books on Architecture, 27 BC). It is particularly famous for asserting that buildings must combine three essential qualities: *firmitas* (solidity), *utilitas* (usefulness) and *venustas* (beauty).

Among Vitruvius's contemporaries was the philologist and polymath Marcus Terentius Varro (116–27 BC) who authored an impressive number of works, among them *Disciplinarum Libri IX* (The Nine Books on Disciplines). These included grammar, rhetoric, logic, arithmetic, geometry, astronomy, musical theory, medicine, and architecture. On this ultimate topic, Vitruvius explicitly mentions Varro as an important reference.

Unfortunately, Varro's corpus (an estimated 74 works in 620 books) is almost entirely lost. His only surviving opus is *Rerum rusticarum libri III* (Agricultural Topics in Three Books), described as "the well digested system of an experienced and successful farmer who has seen and practiced all that he records." Alongside a few others (by Cato, Columella or Palladius), Varro's treatise is one of our best sources of information on the science of farming and the management of rural affairs in classical Rome.

The proximity of these two authors: the architectural theoretician who reflected on the selection and management of sites and the agronomist who also authored a book on architecture, should give us food for thought. Indeed, this intimacy was not lost on the medieval copyists who passed on their works to early Renaissance readers. It so happens that one of the best medieval copies of Vitruvius's *De Architectura*, in the codex *Mediceus Laurentianus*, was bound together with Varro's *Rerum rusticarum* (as well as Cato's *De Agri Cultura*). Apparently, 14th century copyists still held the two disciplines as complementary aspects of the same concern in the management, culture and improvement of the ecumene. Bluntly stated, the ambition of our exhibition is to encourage the rebinding of agriculture and architecture today.

References
Vitruvius, *The Ten Books on Architecture* (trans. Morgan), Harvard University Press, 1914.
Cato and Varro, *On Agriculture* (trans. Hooper and Ash), Harvard University Press, 1934.

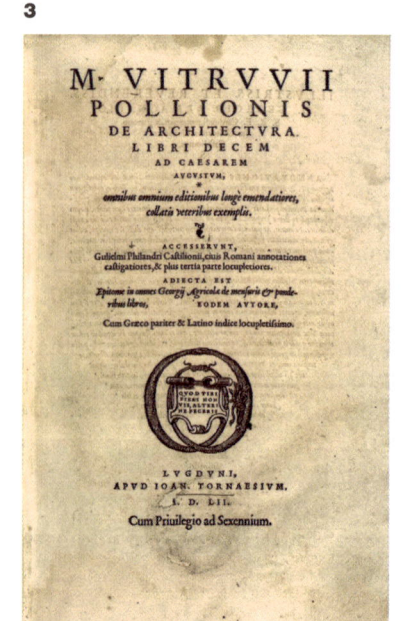

3
Vitruvius's *Ten Books on Architecture*, circa 27 BC.

4
Varro's *Three Books on Agricultural Topics*, circa 40 BC.

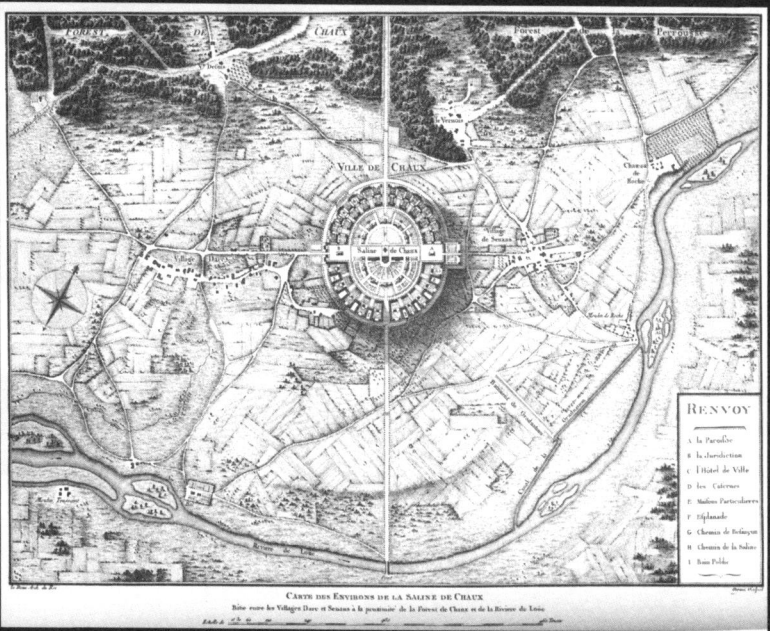

A MANUFACTURED COUNTRYSIDE
Project for the Saline and the city of Chaux, Claude Nicolas Ledoux, 1804.

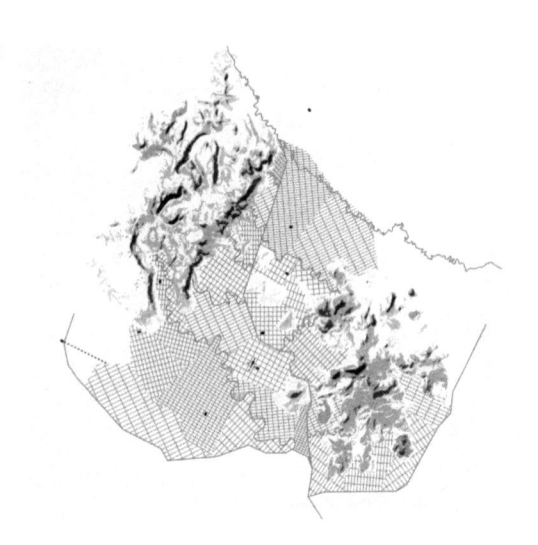

1
Plan of Gallo-Roman villa
in Montmaurin, France,
4th century AD.

2
Map of agricultural lands and
villas near Vicenza, 16th century
(including Palladio's Pisani,
Repeta, Saraceno and Poiana).
Smienk and Niemeijer

VILLA SUBURBANA

DOMINATING THE COUNTRYSIDE

A nodal type in the history of the dialectics between agriculture and architecture in the West is the *villa suburbana*, in which "suburbana" means both out of the city and para- or quasi-urban: an ideal combination of *otium* (cultivated leisure and refinement) and *negotium* (work and commerce), and therefore a close integration of civility and rural affairs.

The Romans famously codified the components of this type along a standard gradient: *pars urbana* (home of the master and his family), *pars rustica* (barns and accommodation of workers or slaves), and *pars fructuaria* (the section of the land that bore fruit), the latter being divided up into *hortus* (garden), *ager* (fields), *saltus* (pastures) and *silva* (forest).

A powerful instrument in Rome's colonization of its empire (a network of latifundias), the villa complex was not just an instrument of domestication but also the seat and symbol of a *domination* of the countryside. And indeed, the revival of this type from the Renaissance on, which substituted the more fortified and walled-up feudal *castrae*, actively accompanied the progressive control of entire estates and territories (both in the visual and political sense, an *integrazione scenica*). The different stages and versions in the evolution of the type are duly recorded in the textbooks describing the history of landscape architecture: the gradual integration of components by Palladio and others in the 16th century; the diffusion of the resulting model through designers like Inigo Jones and Thomas Jefferson; the dramatic expression of political control in the telescopic structures of the classical park in 17th century France; and the partial disintegration of the type (in fact an æsthetic sublimation of subjection) in the English picturesque at the dawn of the industrial revolution.

SÉBASTIEN MAROT

Again, throughout these centuries, the villa suburbana was the theatre and laboratory for regimenting the countryside (a new "nomos of the earth") which developed in both Europe with the *enclosures* and the dismantling of the *commons* and the colonies in the guise of *haciendas*, *plantations*, etc.

References
R. Bentmann & M. Müller, *The Villa as Hegemonic Architecture*, Humanities Press, 1992.
J. Becker & N. Terrenato (eds.), *Roman Republican Villas: Architecture, Context and Ideology*, University of Michigan Press, 2012.

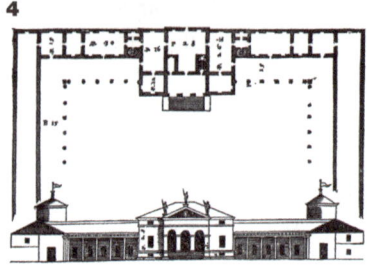

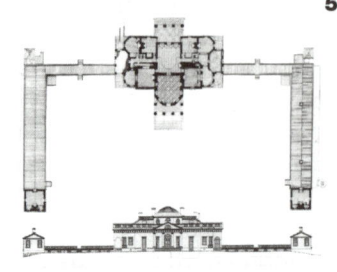

3
Aerial view of Thomas Jefferson's villa and gardens in Monticello, Virginia, 1780–1810.
Phot. Robert Llewellyn, 2010

4
Plan of Villa Saraceno, Andrea Palladio, 1570.

5
Plan of Monticello, Thomas Jefferson, circa 1780.

THREE NATURES (SALTUS, AGER, HORTUS)
Frontispiece of *Curiosités de la nature et de l'art*,
Pierre le Lorrain de Vallemont, 1703.

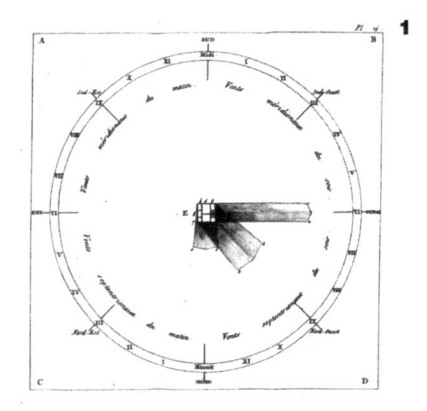

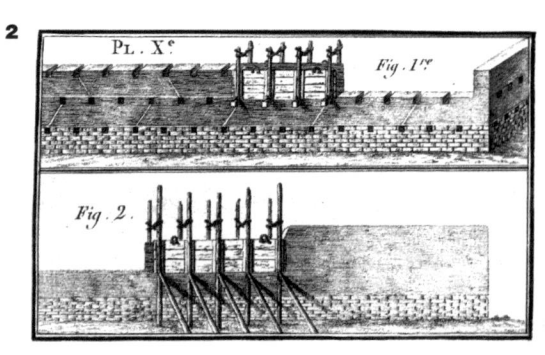

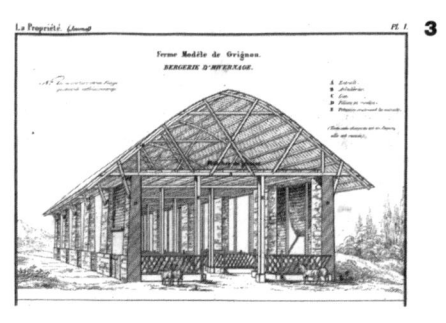

1
Orientation of a house
according to cardinal directions,
François Cointereaux, 1797.

2
Adobe building techniques
for rural architecture,
François Cointereaux, 1790.

3
Sheep Barn, Ferme modèle
de Grignon, 1832.

AGRITECTURE

A PLEA FOR CROSS-FERTILIZATION

> "Architecture is, after Agriculture,
> the first and most useful of arts."
> Claude Jacques Toussaint, 1811.

Shortly before the French Revolution, François Cointereaux, a bricklayer from Lyon, settled in Paris where he gave himself the title of "Professor in Rural Architecture". Until his death in 1830, he promoted adobe or mud-brick masonry techniques in numerous pamphlets and books, but also provided specific models, plans and structures for all kinds of farm yards and rural buildings; all the while suggesting that these could also be applied to urban monuments and cities.

An advocate of rural living, Cointereaux strived to demonstrate economic ways to reunite building and farming skills. Whereas country masons and carpenters are good craftsmen but unable to measure "how the celestial and atmospheric effects must influence the arrangement of our buildings", farmers, although "closer to nature", lack the competence to design the buildings they really need. Cointereaux's ambition was to bridge this gap:

"The public wants a man who can handle both the trowel and hammer, and the spade and hoe. I propose to perform those two functions, and I will show that one can, while improving the land, also build a small house: I will take advantage of the very science of the wine grower, the ploughman and the gardener, to turn them into builders."

Aside from his climatic approach to the design of buildings, and his proposals for optimising their heating and ventilation, Cointereaux was also prolific in inventing original ways of improving lands and their yields, e.g. fences that would fertilise the ground as they degraded, or machines that would help peasants to compress lumps of earth (a bit like adobe) and then disperse them on the fields in order to enrich them.

What Cointereaux strived to promote for over forty years was a cross-fertilisation of architecture and agriculture; a project of osmosis which called for a neologism: "Architecture has always been treated in isolation; Agriculture has always been explained separately. This is a mistake: those two arts won't progress unless one fuses their principles in the same spiritual melting-pot; a new science then emerges, which I, with good reason, name Agritecture." If this neologism failed to be taken up at the time, it is interesting to note that it has recently resurfaced, but with a slightly different agenda: that of incorporating agriculture within metropolises, in the guise of high-tech "buildings that grow food", a considerable simplification of the idea.

References
Jacques Toussaint, *Traité de géométrie et d'architecture*, Paris, 1811.
Jean-Philippe Garric, *Vers une agritecture: architecture et constructions agricoles (1789–1950)*, Mardaga, Bruxelles, 2014.

4
Cowshed, Ferme de Platé, France, 1892.

AGRICULTURAL URBANISM?
Green Belts for London, John Claudius Loudon, 1829.

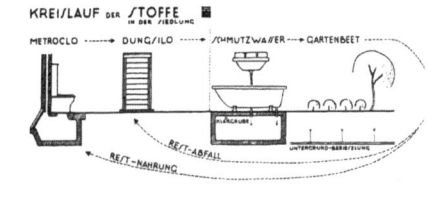

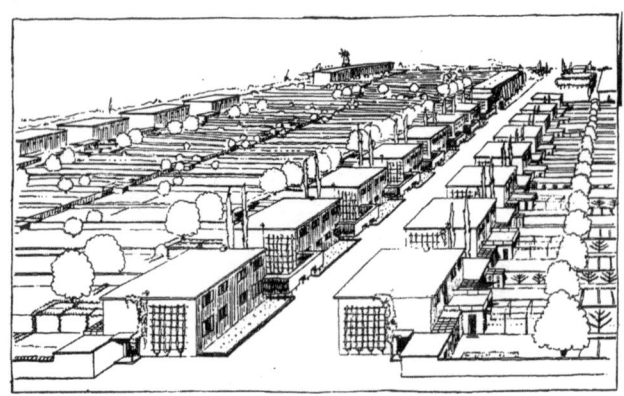

1
Cycle of nitrogen, Raoul Francé, 1922.

2
Migge, Cycle of the elements in Siedlung Ziebigk, 1926.

3
Diagram of a house-garden unit, Adolf Loos, 1920s.

4
Axonometric view of Siedlung Ziebigk, Dessau, L. Fischer and L. Migge, 1926.

SELF-SUFFICIENCY

AN ARCHITECT FOR HORTICULTURE

Leberecht Migge (1881–1935) was a major figure in the Modern Movement and one whose work has been overlooked until recently. A landscape architect influenced by the Garden City and Land Reform movements, Migge promoted an architectonic and functionalist view of garden design which led him to work with key German architects and urbanists such as Wagner, Taut and May. Central to his engagement was a left-wing agenda based on the planning of community settlements (S*iedlungen*), coupled with a biological functionalism focused on the metabolic cycle. At the end of WWI, Migge promoted this practical philosophy of design in two influential texts: *Everyman Self-Sufficient: A solution to the Issue of Community Settlements Through a New Kind of Garden Design* (1918), and *The Green Manifesto* (1919) which branded the term "green" as a label for ecological concerns.

In Migge's view, the productive garden (a combination of open-air rooms) was, with the house itself, an integral part of the dwelling unit. Through the permeable agency of kitchen and glass-housed terrace, where waste (including human's) was turned into compost and fuel, house and garden interpenetrated like two phases in the metabolism needed to keep an organism alive. In the same way that a garden grew but sometimes needed replanting, the house itself was seen as something that could grow or diminish according to needs. In the 1920s, Migge implemented this program in many of the Siedlungen that he planned or built with Wagner, Taut, May or Leopold Fischer around Dessau, Berlin, Frankfurt, etc. as part of the *Binnen Kolonization* (inner colonisation) strategy then deployed in Germany. Beyond the metabolic unit of house and garden, Migge also advocated an analogue approach on the scale of the Siedlungen and cities, including dung factories inspired by the way human waste was systematically collected in Chinese cities and recycled into agriculture.

What is most fascinating is the consistency with which Migge pursued this close combination of technology and biology; of social and ecological concerns in his work, and drew lessons from both history and the most advanced research in pedology, soil microbiology (such as Raoul Francé's *Edaphon*) and gardening techniques.

This practical and theoretical approach culminated in his final opus: *Die Wachsende Siedlung* (The Growing Siedlung, 1932). In this book, the elements of horticulture, such as garden walls which retained solar heat and accommodated trellised plants, fruit trees and houses, became the spines of continuously evolving settlements. Not exactly a "garden-city" perhaps but a genuinely horticultural approach to urban and civic design: the productive garden and its technology as substructure of new settlements. When compared with rural projects advanced by other modern architects of the era, it seems that Migge's work, which anticipated many of the key concerns of permaculture and agroecology, offers a trove of ideas and lessons that could be mined even deeper today.

References
David Haney, *When Modern was Green: Life and Work of Landscape Architect Leberecht Migge*, Routledge, 2010.

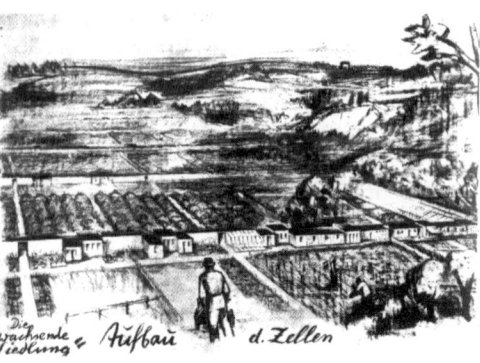

5, 6
Leberecht Migge, *Die Wachsende Siedlung [The Growing Siedlung]*, cover and perspective, 1932.

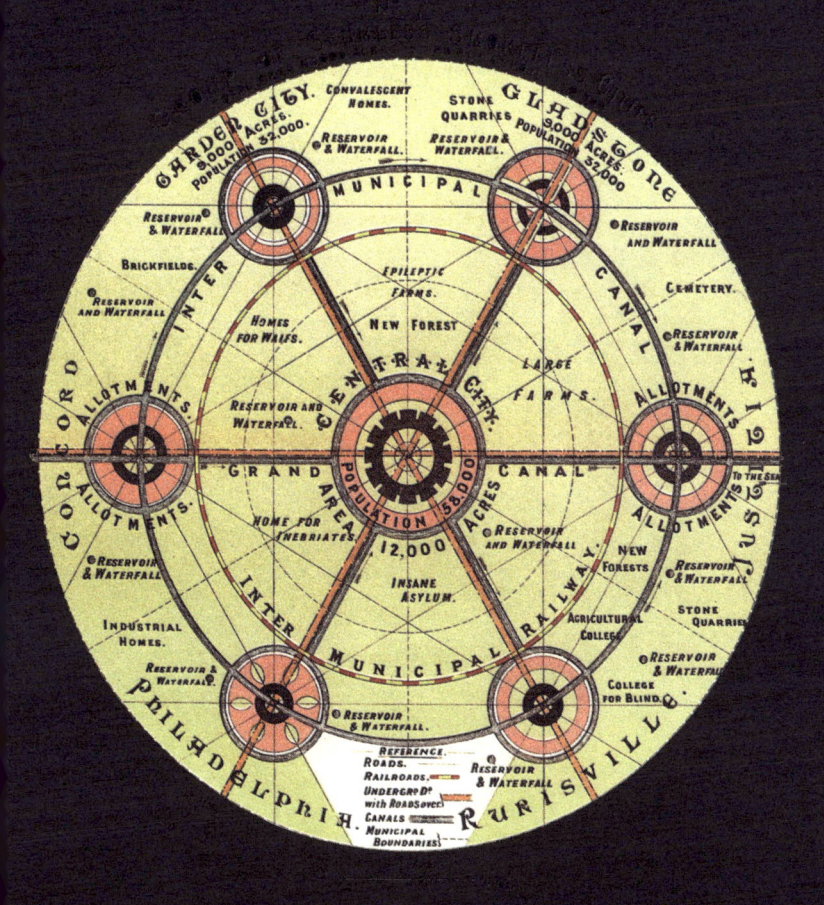

RECONCILIATION?
Ebenezer Howard's Garden City, Diagram no. 7,
To-morrow: A Peaceful Path to Real Reform, 1898.

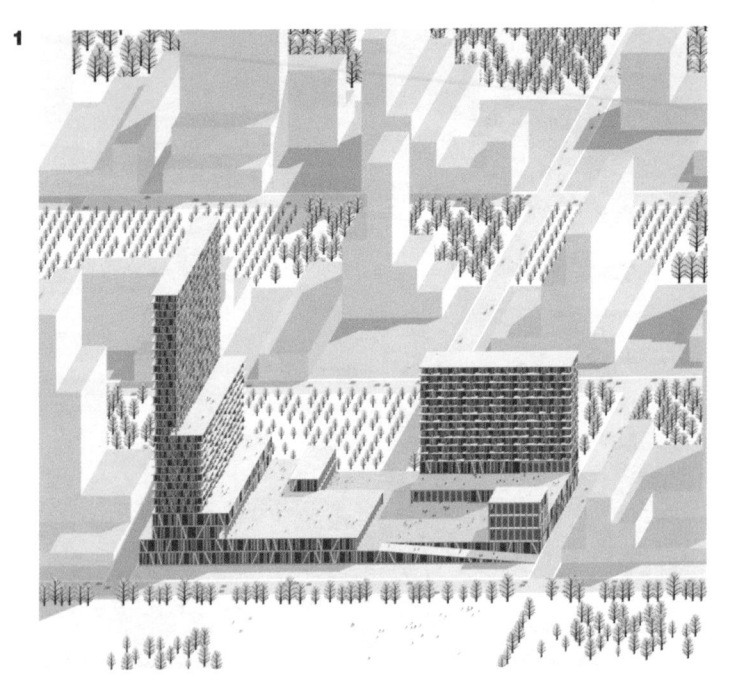

1
Urban block, built out of timber, Albert Pope & Jesus Vassallo, Detroit, 2016.

2
Aerial view of the district, *ibid.*

INTEGRATION?

CARBON 2065

In 1996, architect Albert Pope published *Ladders*, a book which famously identified spine-based developments or "ladders" as the DNA of post-war urbanisation and car-dependent suburbia. Based on cheap fossil energy, these new urban forms rapidly dislocated both the centripetal and centrifugal gridiron of the Open City. In recent years, with colleague architect Jesus Vassallo, Pope developed Carbon 2065, a combined strategy for urban forest plantation and wood construction that would allow cities to become denser while still capturing carbon. The model, which could be realised over the next fifty-year building cycle, is designed to cut per-capita energy consumption in North America by 75%, i.e. the amount required to keep surface temperature warming to below 2 degrees Centigrade. Pope writes: "Built on spine-based urban development, the model combines rewilded green corridors, carbon plantations and cross-laminated timber construction in an attempt to integrate the urban development into the carbon cycle. Beyond its immediate response to the ecological crisis, the project deploys wood in various states in an attempt to create a broad-based material culture capable of replacing vernacular practices that were abandoned over a half-century ago."

The premise of this strategy is that metropolises, although historically the products of the cheap and abundant energy delivered by the massive mining of fossil fuels, could switch to renewable energy and material resources by growing, harvesting and processing wood, without surrendering much of the density, modernity and building-types we have grown used to. Moreover, this integration of silviculture and wood-building industry within the urban fabric would, by verticalising them in high rise, redeem the deleterious effects of these spine-based developments which were the agents of sprawl. Instead of simply being major emitters of CO2, dense cities, where such large wood buildings would be systematically rebuilt every fifty years, would not only reduce global urban emissions but also become active carbon traps.

Of course, some could see this strategy as a consistent attempt at integrating agriculture and urban architecture into the carbon cycle. Others however might stigmatise it as typical of architecture's addiction to the modern metropolis, and as a refusal to come to terms with the most likely territorial consequences of what Odum, Holmgren *et al* call "energy descent.

References
Albert Pope, *Ladders,* Princeton Architectural Press, 1996.

3
Building model, *ibid.*
Phot. Nash Baker

"BUY A COZY COTTAGE IN OUR STEEL CONSTRUCTED CHOICE LOTS, LESS THAN A MILE ABOVE BROADWAY. ONLY TEN MINUTES BY ELEVATOR. ALL THE COMFORTS OF THE COUNTRY WITH NONE OF ITS DISADVANTAGES."—*Celestial Real Estate Company.*

SUPERIMPOSITION
Life, AB Walker, 1909.

B.

AGRI CULTURE & URBAN ISM

In which the intrigued reader sees how cities themselves proceeded from agriculture and its intensification; how the hierarchy of the most needed crops for food, fibres and energy shaped their surrounding landscapes until the Industrial Revolution; and how urbanism then largely prospered on an evacuation of the countryside.

1
Remains of what is often
called the first city: Uruk,
Southern Mesopotamia.
Phot. Essam al-Sudani

2
Intensification: water management
and agriculture, Southern Iraq.
NASA, from *The Ancient Near East*,
Liverani, 2014

3
Aerial View of Arslantepe,
Eastern Anatolia.
Italian Archeological Mission
in Eastern Anatolia, from Liverani,
ibid.

THE URBAN REVOLUTION

COMPLEXIFICATION / SIMPLIFICATION

Around the major areas of the world which saw the emergence of a Neolithic domus complex (The Fertile Crescent, Northern China, the Andes and Mexico), it took several millennia before this complex started to reach the size and proportion of a city, with several thousand inhabitants, public buildings, monuments and infrastructure. In most cases (with the notable exception of the Andes), this "urban revolution" (as Gordon Childe named it) took place in the naturally rich ecosystems of large river valleys and marshlands, regularly visited by fertilising floods, and where, for that very reason, people had not initially resorted to strictly agricultural practices: Lower Mesopotamia, The Nile and the Indus valleys, the Yellow River, the Aztec Marshlands, etc.

In these areas, the emergence of cities corresponded to the interlocking of several major agricultural innovations such as water management infrastructures (for drainage more than irrigation), the corresponding geometric division of fields, the recourse to draft animals and machinery (for ploughing and seeding) and a systematic focus on annual cereal crops such as wheat, rice and corn, which were visible, seasonal and thus easy to be taxed. Obviously, society grew more complex in these first cities. Labour was divided. Social hierarchies were created. Elaborate frameworks of organisation and administration, assisted by the invention and development of writing, came into being. These went hand in hand with a no less considerable simplification of ecosystems, focused on extracting surpluses by extending the control and selection of energy converters such as plants, animals and humans (notably in the form of slavery).

As incubators of the transformation from chiefdom to state and eventually empires, these first cities were a stepping stone along the path that led from the neolithic multispecies resettlement camp to the human park of modern nations and their global biopolitics.

Since the onset of the Industrial Era, and the massive rural exodus it provoked, we have tended to forget that double bind (complexification / simplification) as if it had been overcome by what we think of as progress. But one should keep in mind that, far from being a continuous success story, the evolution of the world since the neolithic time has also been a long chronicle of recessions and collapses of complex societies, all more or less directly caused by the way they mismanaged their resources and limits to growth.

References
Vere Gordon Childe, "The Urban Revolution", 1950.
Peter Sloterdijk, *Regeln für den Menschenpark*, Suhrkamp, 1999.
Mario Liverani, *Uruk, The First City*, Equinox Publishing, 2006.
Mario Liverani, *The Ancient Near East: History, Society and Economy*, Routledge, 2014.
James C. Scott, *Against the Grain: A Deep History of the Earliest States*, Yale, 2017.

4
Architecture, garden and water:
Ipuy tomb at Deir-el-Medineh,
Egypt, circa 1250 BC.
Facsimile Norman de Garis Davies,
1924

IDEAL TERRITORY / ULTIMATE REFINEMENT
Garden Carpet, Iran, early 18th century.

SÉBASTIEN MAROT

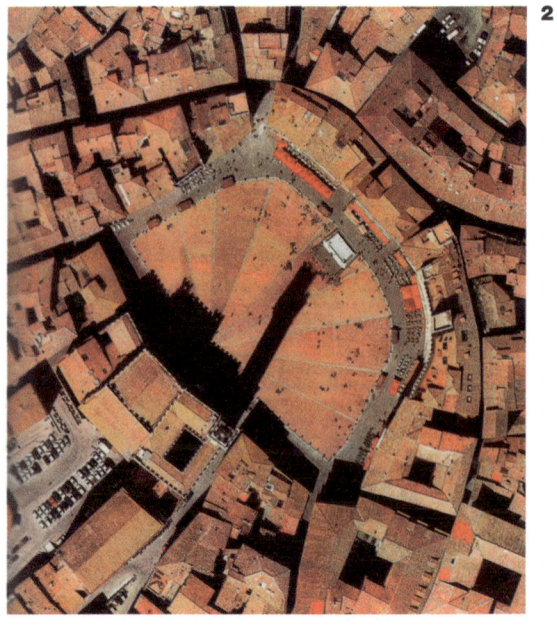

1
Umbilicus Urbis: Mundus in Ostia Antica, Italy.
Roger Ulrich, 1985

2
Sienna's Piazza del Campo, Italy, with Gaïa Fountain and water drains.
Patrick Boucheron, *Conjurer la peur,* Seuil Paris 2013

MUNDUS

THE COSMIC
VERSUS GLOBALISM

Much has been written on the rituals in the foundation of ancient cities, particularly in classical Rome: the layout of the *cardo* and *decumanus;* the delimiting of the *templum* area; the digging of the *pomerium* ditch with a plow pulled by oxen etc. But little attention was paid until Joseph Rykwert's *The Idea of a Town: The Anthropology of Urban Form in Rome, Italy and the Ancient World* to the ritual of the *mundus*, which consisted in the digging of a large central pit, complementary to the temple, into which colonists threw handfuls of earth from their places of origin, and where citizens regularly threw fruits from their fields that surrounded the city. Most significantly, it was during the inauguration of this *mundus* (called the *umbilicus urbis*) that the future city was given its name.

The *cardo* and *decumanus* crossed at right-angle and organised the plan of the city. But this two-dimensional plan was completed by a vertical axis symbolised by the *mundus*. Each city or town was thus rooted in and around an allegory of the productive landscape that fed it, a rounded cavity which mirrored the vaulting sky and gave the city the dimensions of a volume, or kosmos: a world within the world.

One might wonder if the *mundus* was not the ancestor, together with the *agora*, of large public *campi*, such as the stunning Piazza del Campo in Sienna, created in the 14th century, which looks indeed like the symbol and miniature of the city's cultivated landscape: a structured and evenly-parcelled watershed, complete with Gaia fountain and converging drains. Furthermore, one might also wonder if Central Park, and Koolhaas's *City of the Captive Globe*, should not be read as late heirs of that legacy, and symbols of the progressive globalisation of the *mundus* in the modern metropolis. A vertiginous acceleration which raises the following question: to what extent can the *mundus* be enlarged and extrapolated without losing the qualities of the ancient Greek meaning of cosmic: namely roundedness and resilience? Or, put differently: can one really pretend to be a citizen of *the* world, unless one is actively immersed in the building and careful maintenance of *a* world as its willful inmate or member?

References
Joseph Rykvert, *The Idea of a Town: The Anthropology of Urban Form in Rome, Italy and the Ancient World,* Faber & Faber, 1976.

3
Central Park, New York, aerial view, 1857–1876.

4
Rem Koolhaas and Zoe Zenghelis, ***The City of the Captive Globe,* 1976.**

"CULTURAL CLEARING"
From *Il paesaggio come teatro*, Eugenio Turri, 1998.

1 2

AGRICULTURE AND ARCHITECTURE

IMAGO MUNDI

"WARDING OFF FEAR": METASTASIS AND RUIN

A commission of the rulers of Sienna, the Effects of the Good and Bad Government frescoes were painted around 1338 by Ambrogio Lorenzetti, just after the completion of Piazza del Campo. More than any other artwork, those two opposite cross-sections through city and country express what Georg Simmel would write in "The Sociology of Space" (1903): "The picture frame, the boundary enclosing an entity, has a very similar sense for a social group and for a work of art... Delimiting and collecting it against the world that surrounds it; the picture frame claims that within it is a world that only obeys its own rules."

A powerful memento of the consequences of the ruler's good or bad morals and policies, the two frescoes show that the cityside and countryside of each world are fully interdependent and united for the better, and for the worse. Whereas the Good Government depicts a peaceful flow of goods and people between a lively city (teeming with activities amidst well maintained buildings) and a palette of self-sustaining polycultural practices, linked to the larger world by the umbilical cord of a river meandering toward a sea-port on the far right, the Bad Government, itself badly injured, features the pending evils (desolation, war, crime and ruin) of unwise management and unsustainability.

In other words, Lorenzetti represents the world as a relatively cohesive *regio* where the city, clearly core and collector, nonetheless constitutes only one sphere or half of the picture, integral with its rural counterpart, the well-tended mosaic of cultures which provides for most of its basic needs. If the fortified wall that encloses the city, and cuts the image in two halves, suggests that the world of this commune might provisionally retreat behind it to find shelter in times of crisis, the opposite fresco warns that this enclosure will be no rampart against the deleterious effects of war and political neglect.

The painting does not just portray the world as an "anthropogenic island", as Peter Sloterdijk has it, or as a humanised clearing, but as the ideal of a country whose works and days have turned into a metastable milieu, the operating theatre for a collective life and peaceful cohabitation. Indeed, what Lorenzetti depicts in this anamorphic section are the body and organs of such a world, the gradient of a dia-cosmos, a spoonful of landscape and activities, stretched between the volumetric beehive of the city and the promise of the elsewhere featured by the distant port towards which the river meanders.

References
Emilio Sereni, *Storia del paesaggio agrario italiano*, Laterza, 1961.
Patrick Boucheron, *Conjurer la peur, Sienne 1338, Essai sur la force politique des images*, Seuil, 2013.

1
The Effects of the Good Government, Ambrogio Lorenzetti, Palazzo Civile, Sienna, 1338.

2
The Effects and Allegory of the Bad Government, Ambrogio Lorenzetti, Palazzo Civile, Sienna, 1338.

FROM GREENHOUSE TO CRYSTAL PALACE
Great Exhibition, Joseph Paxton, 1851.

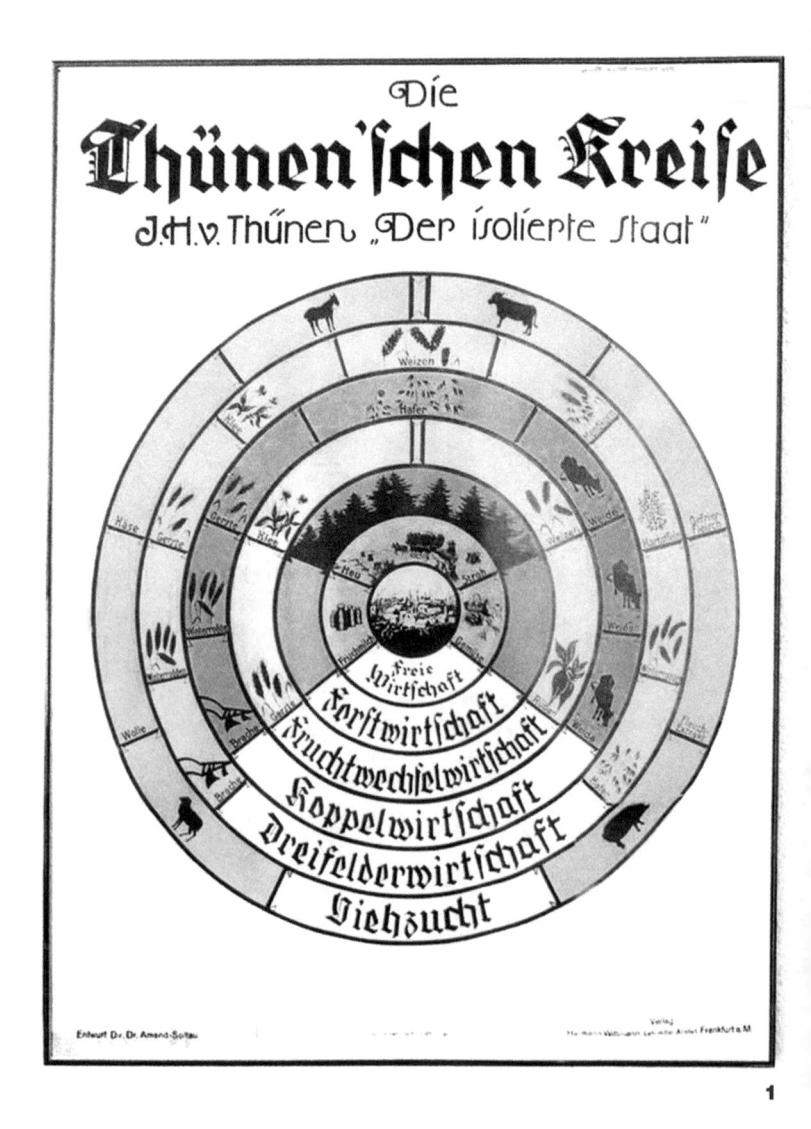

1
Johann Heinrich von Thünen's diagram of the isolated state and its rings of crops and resources, 1826. Redrawn in 1920s for use by schools.

AGRICULTURE AND ARCHITECTURE

LAND USE MODEL

THE GRADIENT OF PRE-INDUSTRIAL LANDSCAPES

In 1826, the German landowner and economist J. H. von Thünen published an essay entitled *Der Isolierte Staat in Beziehung auf Landwirtschaft und Nationalökonomie*, where he proposed a theoretical model of the way an "isolated state" would naturally tend to use its land. As Carolyn Steel writes in her book *Hungry City*: "The eponymous 'state' consisted of a 'very large town' in the middle of a featureless, fertile plain, the latter inhabited only by rational, profit-seeking farmers. Under such conditions... the farm belt around a city would organize itself into a series of concentric rings, like ripples from a pebble chucked in a pond. The innermost one would consist of market gardens and dairies, whose profits would most benefit from the city's manure. Beyond this would be a ring of coppices for firewood; and beyond that, arable land where the city's grain would be grown, close enough to the city to make its transport feasible. Beyond that, there would be grazing for livestock, and finally, wilderness: land so far from the city that it wasn't worth exploiting. (The question of where the inhabitants of the Isolated State might want to spend their weekends did not arise)."

In other words, von Thünen saw the economy of preindustrial territories as basically structured by the more or less profitable, transportable or perishable quality of basic resources (food, fibres or energy) sustaining the central city where they converged. Of course, his model, projected on a clean slate, rarely corresponded to the situation of actual regions where the contingencies of geography, the presence of roads or other cities significantly distorted this simple geometry. But provided one took these contingencies into account, von Thünen's model effectively depicted the basic parameters and gradient of land-use in a mostly agrarian economy.

As Steel says: "Despite its disregard of politics, culture and most forms of geography, von Thünen's land-use model mirrored the pre-industrial world rather well. Of course, it was essentially a mathematical rationalization of the way most cities had been fed up until his day — cities, that is, without recourse to the most important influence on food supplies in the pre-industrial world: the sea."

Indeed, due to comparative advantage of shipping for access and supply, the sea and large rivers had already wrenched certain cities and regions away from the *nomos* of the local and the metabolism of the isolated state, and turned them into cosmopolises, long distance magnets to resources and capital; where another way of being in the world progressively evolved. But it was really with the great transformation brought about by the industrial revolution, which was fundamentally a revolution in energy, motors, and transportation, that the model of the isolated state — at the very moment when von Thünen formulated it — began to dissolve.

References
J. H. von Thünen, *Der Isolierte Staat in Beziehung auf Landwirtschaft und Nationalökonomie*, 1826.
Carolyn Steel, *Hungry City: How Food Shapes Our Lives,* Vintage, 2008.

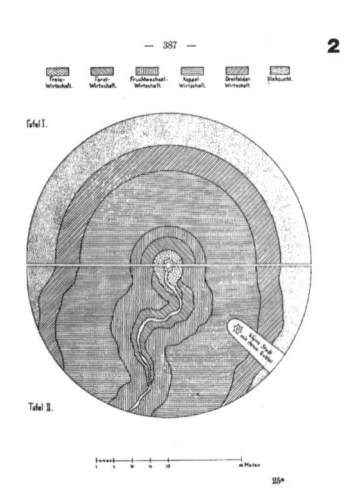

2
Diagram showing the effect of a river on von Thünen's land use model, 1850.

VvR 3250

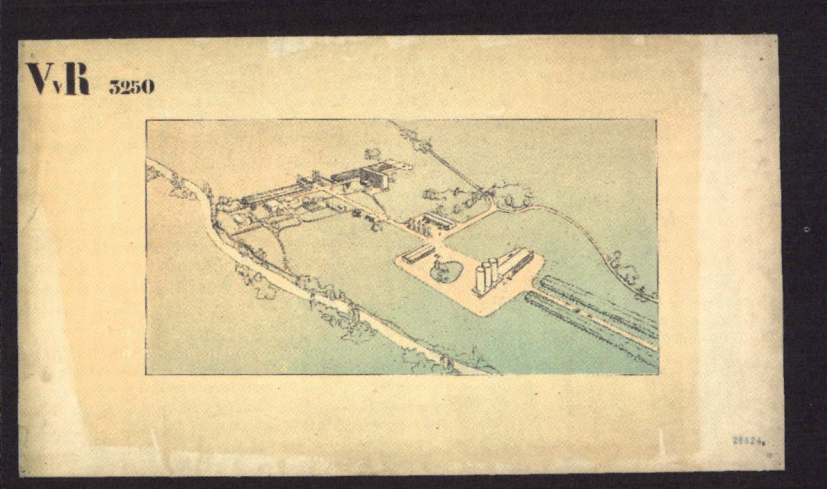

RADIANT FARM AND COOPERATIVE VILLAGE
Le Corbusier, 1939.

1
The world's railroad scene:
Illinois Central Rail Road
Charles I. Felthousen, 1882

2
Disassembly Line: Armour
Company's pig slaughterhouse,
Chicago, 1891.

NATURE TO MARKET

"ALL THAT IS SOLID MELTS INTO AIR"

"If the frame structure is the essence of modern architecture then we can only assume a relationship between ourselves and Chicago comparable to that of High Renaissance architects with Florence, or of the High Gothic architects to the Ile-de-France. For, although the steel frame did make occasional undisguised appearances elsewhere, it was in Chicago that its formal results were most rapidly elucidated."
Colin Rowe, "The Chicago Frame", 1956.

If one were to select one metropolis as the capital of modern architecture, Chicago would be the most obvious contender. And indeed, this tiny village (30 houses in 1830) which grew into the 4th largest metropolis in the world by the end of the 19th century, could also be seen as a major symbol and laboratory of the modern city brought about by the industrial era. Alvin Boyarsky pointedly demonstrated this in *"Chicago à la Carte*: The City as an Energy System" (1970), a profusely illustrated essay which paved the way for Rem Koolhaas's *Delirious New York* (1978). But these architects only captured one half of the picture.

The most instructive book ever written on Chicago is probably William Cronon's *Nature's Metropolis* (Norton, 1991), which basically portrayed this megapolis as a product and motor for the industrialisation of agriculture in the Midwest. Cronon minutely documented how the city, thanks to the spread of technical networks such as the railway and the telegraph, and innovations like steam-powered grain elevators, entirely reorganized the production, transformation and market of basic resources (grain, lumber and meat) over an immense territory whose ecosystem was radically transformed in the process: The effect was huge. While native cereals and leguminous plants were substituted for corn and wheat, and slaughtered herds of bisons by ranches of imported pigs and longhorns — which would all end up in the disassembly lines of Chicago's meat district —, the huge forests of white pine of Illinois and Michigan were literally mined and drained (via skid rows, rivers and railways) to the lumberyards of the city.

"From the wealth of nature," Cronon writes, "Americans had wrung a human plenty, and from that plenty they had built the city of Chicago. Chicago's relationship to the white pines had been exceedingly intricate, emerging from ecological and economic forces that for a brief time had come together into a single market, a single geography. The tensions in that market and in that geography finally destroyed the distant ecosystem which had helped create them — but then it no longer mattered. Perhaps the greatest irony was that by surviving the forests that had nurtured its growth, Chicago could all too easily come to seem a wholly human creation."

References
Colin Rowe, "The Chicago Frame", *Architectural Review,* 1956.
Alvin Boyarsky, "Chicago à la Carte: The City as an Energy System", *Architectural Design*, Dec. 1970.
William Cronon, *Nature's Metropolis : Chicago and the Great West*, Norton, 1991.

3
Chicago's lumber district.
Charles Graham, 1886

SHOPPING MALL
Northland Center, Victor Gruen, 1954.

SÉBASTIEN MAROT

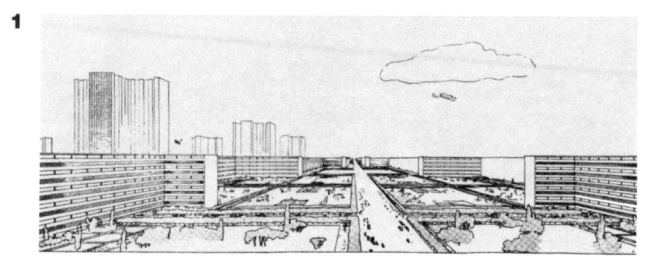

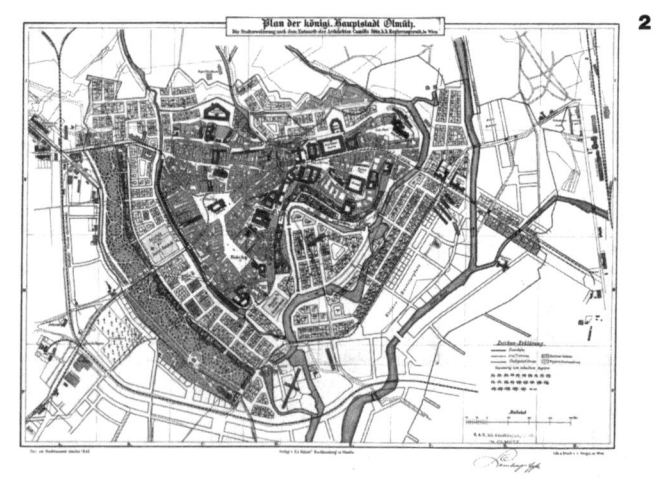

1
Progressism: Ville contemporaine pour trois millions d'habitants, Le Corbusier, 1922.

2
Culturalism: Extension Plan for the City of Olmütz, Camillo Sitte, 1895.

3
Naturalism: Emerald Necklace and Boston Park System, Frederick Law Olmsted, 1894.

66

AGRICULTURE AND ARCHITECTURE

URBANISM

THE CONFLICTING TRENDS
AND SCHOOLS OF A NEW DISCIPLINE

The term "urbanism", which became common currency in the first decade of the 20th century, was in fact coined 50 years earlier by designers, engineers and administrators such as Cerda and Haussmann, and substituted antiquated expressions such as "urban embellishments" or "urban art". If one pays attention to the historical context in which it emerged, however, the term does not designate the art of building cities as much as that of managing the explosive growth of extant cities due to the industrial revolution and the influx of rural migrants it provoked. Taking stock of a century of urbanistic literature, historian Françoise Choay distinguished three major competing trends in its jurisprudence. Using her denominations, one might characterize them as follows:

A. Progressism was the common ideology of those who insisted this explosive growth amounted to a change in nature. What was needed was a thorough reframing of the city based on an analytical understanding of the latent programme of the modern city, its new technologies, ways of living and so on, and on segregating its functions for increased efficiency. It sums up the modernist agenda, and the Charte d'Athènes in 1933 was its clearest and most expressive formulation.

B. Culturalism was the philosophy of those who held that for all their magnitude, which had to be controlled anyway, urban extensions were still proceeding from an extant organism that had developed its forms and syntax over time. The task of urbanism was thus to understand that organism, its DNA and patterns so as to transpose and adapt them. Camillo Sitte's *City Planning According to Artistic Principles* (1889) was the textbook of this approach.

C. Naturalism was the alternative option which held that the main guide and inspiration for managing urban extensions had to be the very territory on which they were deployed:

the environs or surrounding landscapes of the city, their topography, pedology and hydrography were to be the guidelines, backbones and infrastructures of a landscape urbanism. Much more than Frank Lloyd Wright and his Broadacre City (the sole example cited by Choay), one may consider Frederick Law Olmsted and his park systems as the genuine champion of that philosophy.

In order to complete Choay's list we might add Regionalism even though it sometimes converges with both Culturalism and Naturalism. Unlike the other categories however, Regionalism postulated that the city was only one segment of a larger whole which comprised agriculture as well as all the activities that sustained it. This larger whole, the region, was the real subject of that discipline which Patrick Geddes called "geotechnics" or "applied geography": applied, that is, "to the goal of making the Earth more habitable".

References
Françoise Choay, *L'Urbanisme, utopies et réalités, une anthologie*, Seuil, 1965.

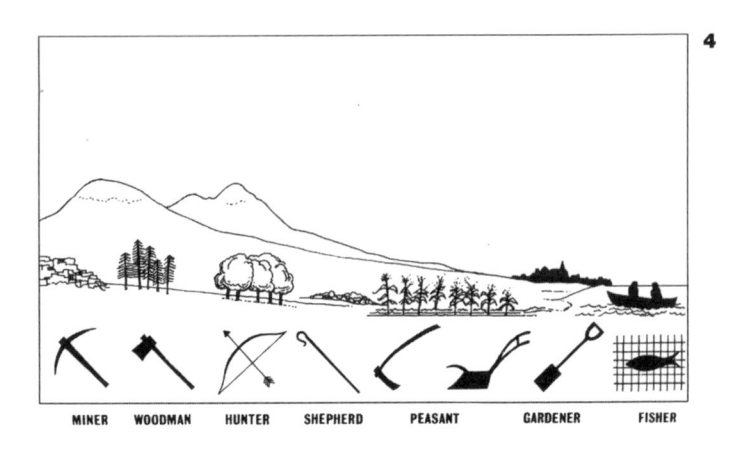

4
Regionalism: Valley section,
Patrick Geddes, 1909.

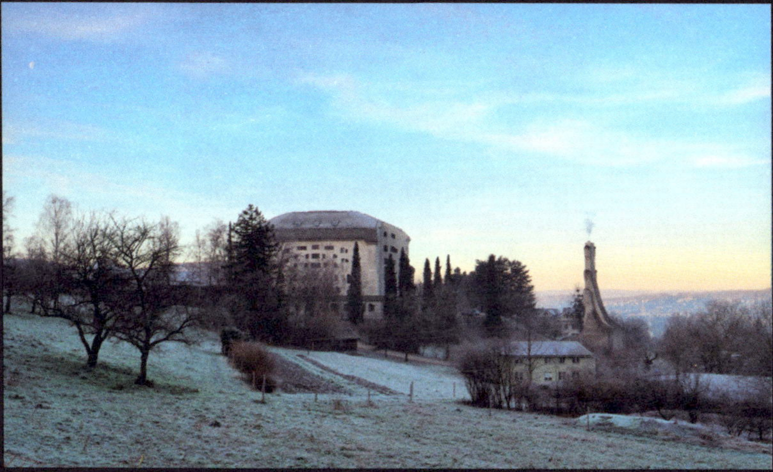

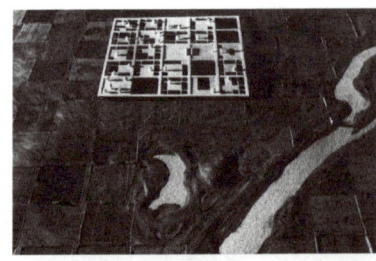

1, 2
New Babylon, Constant
Nieuwenhuis, Netherlands,
1959–74.

3, 4
Agricultural City, Kisho Kurokawa,
Japan, 1960.

AGRARIAN URBANISM?

THE OU-TOPIAS OF SUPERIMPOSITION AND HYBRIDIZATION

After Howard's Garden City, Wright's Broadacre or Hilberseimer's New Regional Pattern, architects occasionally came up with urban visions which attempted to overcome the binary opposition of city and country, and instead produce what Waldheim called "a commingling of the agrarian and the urban". The examples shown here represent three different poetics or design strategies in this respect.

Constant's New Babylon, a reticulated collage of urban plans, features a nomadic network of megastructural corridors spreading out above the earth like a net. It delimits a mosaic of countrysides that assume in his drawings the benign outlook of picturesque meadows and openfields. Consistent with Constant's hypothesis of Homo Ludens being the expected outcome of the machine age, work is conspicuously absent from those images.

Designed after the dramatic flood caused by the Ise Bay Typhoon in 1959, Kisho Kurokawa's metabolist project for an Agricultural City offers the model of an urban mat but one susceptible to variation and proliferation, raised on stilts above the geometrical checkerboard of rice paddies. An obvious solution to the flooding of buildings, but slightly puzzling in terms of agronomic efficiency.

Finally, Andrea Branzi's Agronica (a contraction of *agricoltura* and *elettronica*) explored the "weak urbanism" that might evolve from the hybridisation of urban practices and buildings, partly rendered fluid, nomad, and recombinant by computers, with the seasonality, reversibility and porosity of agricultural fields and activities. An evenly distributed smart grid of "agricultural posts", servicing all kind of activities, constitutes the basic fabric between, as Branzi puts it, "the woof of the artificial and the warp of the natural".

Although one notices an increasing role given to agriculture and agricultural practices in shaping a hybrid rural and urban environment, each one of these systematised visions still considers the farmed, rural world as basically a backdrop or carpet onto which the urban components — be they solid or transient — act as three-dimensional protagonists.

References
Mark Wigley, *Constant's New Babylon: The Hyper-Architecture of Desire*, 010, 1998.
Kisho Kurokawa, *Metabolism in Architecture*, Westview Press, 1977
Andrea Branzi, *Weak and Diffuse Modernity*, Skira, 2006.
Charles Waldheim, "Agrarian Urbanism and the Aerial Subject", chapter 7 *of Landscape as Urbanism*, Princeton, 2016.

5, 6
Agronica, Andrea Branzi, Italy, 1993–94.

"WHAT MUST WE DO TO STOP THE MACHINE SEIZING UP?"
Sicco Mansholt, letter to Franco Maria Malfatti, President of the European Commission, Brussels, 9 February, 1972.

C.
FROM AGRONOMY TO AGROECOLOGY

In which readers are treated first to a survey of the rise and devolution of agronomy over the past four of five centuries, then are granted an overview of the dire consequences of the industrialisation of agriculture and its worldwide export (the so-called "Green revolution"), and finally are introduced to a small compendium of lessons.

1, 2
Plowing and pasture scenes from
Théâtre d'Agriculture, Olivier de
Serres, 1600.

A HOBBY FOR URBAN ELITES

THE SLOW RISE OF AGRICULTURE AS AN ART

For many centuries, agriculture was by far the major source of wealth, even for the aristocracy. Of course, the elites craved expensive hand-crafted objects, often imported from afar, items like jewelry, glass, pottery and fabric. But most of these were paid with income drawn from peasant communities, which formed the great majority of the people. Peasant labour was the basis of wealth accumulation, and the control and command of this labour was the major axis around which political struggle and war were waged. The church may have celebrated the role of agriculture and bookkeeping may have given a prominent place to farm products but farmers themselves were at the bottom of society, just above slaves. As a consequence, farming was regarded poorly and the elite granted agronomy very limited intellectual gravitas or time. The only exception was maybe gardening. Designing and managing a garden was considered noble enough to allow members of the aristocracy to practice it without losing their dignity.

As cities prospered, the growing wealth and political power of craftsmen and tradesmen challenged the dominating landowners. Facing the double threat of merchants and craft guilds to their social and political status, landlords paid more attention to the management of their rural estates. It is remarkable, for example, that it was in merchant city-states like Athens, Carthage or Rome that the first treaties on *agronomy* were produced, one of the most famous being Mago's which is sadly lost to us. These treaties were mainly management handbooks for landowners, embracing the choice of site and soils, servants, slaves, the building of barn and house, the role of the wife, the relationship with the neighborhood, the feeding of livestock, the care of wine and crops and other such pressing managerial matters. The *Res Rustica* by Columella, a Roman landowner from southern Spain (1st century AD), is a perfect example of this literature which encouraged landlords to improve their properties.

Significantly, those writers often complained that both landlords and local authorities did not afford farming the correct respect as a practice.

During the Renaissance, the printing press facilitated the rediscovery of those roman agronomists and inspired a new generation of authors, such as Olivier de Serre, who progressively developed a more experimental approach, and above all convinced themselves that yields could be improved by both research and investment. In the mid-18th century, the physiocrats insisted that states needed to encourage investment in agriculture and called for a reduction in the level of taxes. This political claim was the product of a profound shift in thinking amongst the urban elite: no more considered as a pure source of income, agriculture was to be regarded as an "art".

References
Martine Gorrichon, *Les travaux et les jours à Rome et dans l'ancienne France,* Université de Tour, 1976.
Mathieu Arnoux, *Le temps des laboureurs,* Albin Michel, 2012.
Christophe Badel & Hervé Inglebert, *Grand Atlas de l'antiquité romaine,* Autrement, 2014.

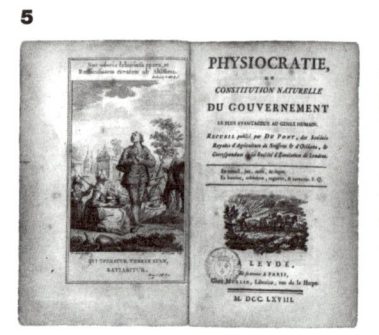

3, 4, 5
Treatises, from agronomy to political economy: Columella's *De Re Rustica* (1st century AD), Olivier de Serres's *Théâtre d'Agriculture* (1600) and François Quesnay's *Physiocratie*, 1768.

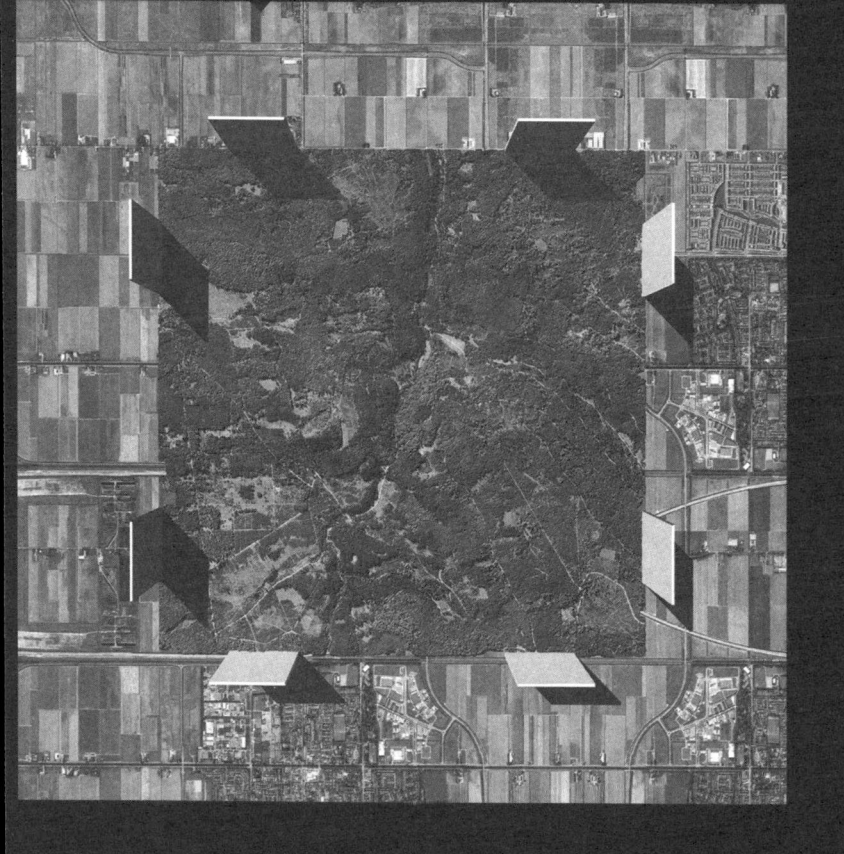

SECESSION?
Stop City, Pier Vittorio Aureli & Martino Tattara, 2007.

SÉBASTIEN MAROT

1
A sugar cane plantation in the French West Indies, 1780.

2
Enclosures: Reparcelling of lands on a plan of the Baronnie of Rots, France, 1666

AGRICULTURE AND ARCHITECTURE

PLANTATIONS AND ENCLOSURES

FEUDALISM VERSUS LATIFUNDIA

In the 3rd century AD, largely due to the diminution in power of the Roman state and the falling off of long distance trade, landowners of the late Roman empire progressively evolved into lords and barons controlling both justice, military power and the economic management of the population. The lack of tradable currencies and the decrease in exchanges led to a more autarkic economy where the lord's power was balanced by a whole set of local rights. Absolute ownership over land and humans (*usus et abusus*) decreased in favour of a complex system of rights and duties, the "*custom*", which significantly rebalanced the power towards peasant communities between the 10th and the 14th centuries. Contrary to common belief, these peasant communities and their villages, where as much seats of genuine democracy (if not more) as the cities. This socio-political system (and its mosaic of *commons*), known as feudalism, was to be dismantled in England and the Netherlands around the 17th century, but persisted in most of Europe until at least the end of the 18th century.

In 15th century Spain, the *Reconquista,* soon followed by the conquest of South America, gave rise, in connection with the redevelopment of international trade, to a new class of huge export-oriented landowners. The extensive cultivation of a few crops on those *plantations*, especially sugar cane and tobacco, spurred the reinstatement of slavery and thereafter the Atlantic slave trade. In England, feudal lords increasingly viewed themselves as landowners with absolute ownership of the land, and no longer as community leaders with a duty of care to the rural population. With the active support of the wealthiest peasant families, eager to incorporate and consolidate their shares of the commons, they abolished the "custom" to develop extensive and export-oriented agriculture, especially for wool. This evolution known as the *enclosures* — plots were fenced against trespassing — deprived small peasants of their access to land. Unable to subsist on their small plots, the poorest farmers moved either to the towns to become workers for the growing industry or to the colonies. North western Europe's early urbanisation also created a market for the products of landlords in Eastern Europe, essentially for grains, which aggravated serfdom in Russia, Poland or Prussia. The result was the compelling decline of free peasantry and small-scale farming, and a growing specialisation in particular regions which later paved the way for industrialisation.

References
Karl Polanyi, *The Great Transformation,* Farrar & Rinehart, 1944.
Femme S. Gaastra, *The Dutch East India Company: expansion and decline,* Walburg Pers, 2003.
Benoit Daviron & Gilles Allaire, *Energie, biomasse, hégémonie: une histoire longue des transformations des agricultures* in *Transformations agricoles et alimentaires entre écologie et capitalisme,* Quae, 2017.

3
Surveyor at work on enclosure at Henlow, Bedford, 1798.

PHALANSTERY
Vue d'un phalanstère, village français, Charles-François Daubigny, 1847.

SÉBASTIEN MAROT

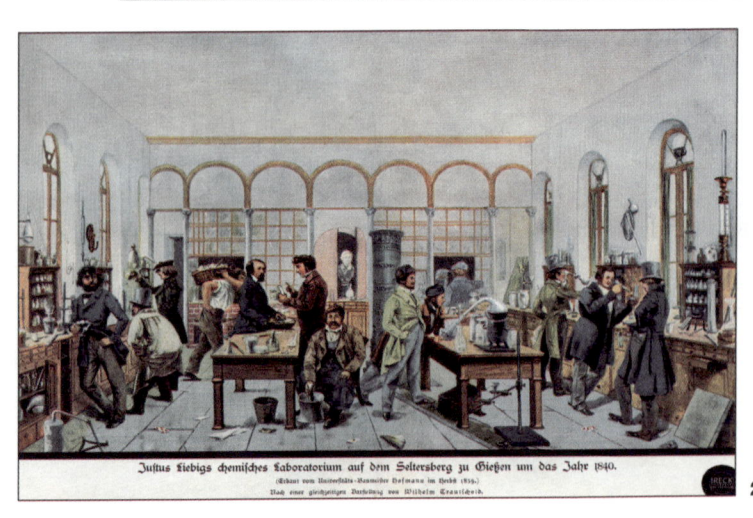

1
Lavoisier demonstrating his discovery of oxygen in 1776.
Louis Figuier, 1874.

2
Justus Liebig's laboratory of chemistry.
Wilhelm Trautshold, 1840.

AGRICULTURE AND ARCHITECTURE

FROM ALCHEMY TO CHEMISTRY

THE QUEST FOR THE ELEMENTS OF GROWTH

In the pre-modern era, the dominant idea was that the material world consisted of four elements: Earth, air, water and fire. Alchemists though kept investigating the true components of matter and their possible *transmutation*. Despite many mistakes and false theories, this quest bore fruits and finally gave birth to the field of chemistry in the 17th century. It wasn't however until 1869 that Mendeleïev published his periodic table of elements (or atoms). Through their investigations into the actual components of matter, alchemists and their successors hoped they would master Nature, and, to some extent, they succeeded. In the history of chemistry's compelling rise we may discern a few milestones or turning points particularly with regard to Nature:

Antoine Lavoisier (1743–1794) established the law of mass conservation (*"Nothing is lost, nothing is created, everything is transformed"*). It was the major outcome of a whole string of enquiries into fire and the burning process. Comparing his own results with those of other scientists, Lavoisier discovered that burning was in fact an interaction of combustible material with an element in the surrounding air which he determined to be oxygen. This led to a deeper understanding of the exchange between living bodies and the atmosphere, and was therefore a major step in understanding respiration and, later, photosynthesis. However, Lavoisier himself did not make a practical connection between his work and agriculture, even though he also paid attention to it. In fact, he wrote a note for the revolutionary government of France about the issue of wheat supply.

Justus von Liebig (1803–1873) may be considered the father of organic chemistry: the chemistry of living beings. He focused his work on the molecules (sets of atoms) which constitute living bodies, and established the basis for the classification of food components (lipids, proteins, glucids and so on). He also demonstrated that it was possible to grow plants without soil, with only a solution of minerals, thus determining the major role of nitrates, without exactly knowing their origin. Later, Hermann Hellriegel (1831–1895) demonstrated the capacity of some plants — particularly the leguminous — to capture nitrogen and enrich the soil in nitrates. Martinus Willem Beijerinck (1851–1931) subsequently determined that the real actors of nitrogen capture were microorganisms present both in soil and in symbiosis with the leguminosa.

If alchemists and chemists failed to turn lead into gold, they succeeded in understanding how life turns air into organic matter.

References
Paul Robin, Jean-Paul Aeschlimann & Christian Feller (eds.), *Histoire et agronomie: entre rupture et durée,* IRD Éditions, 2007.

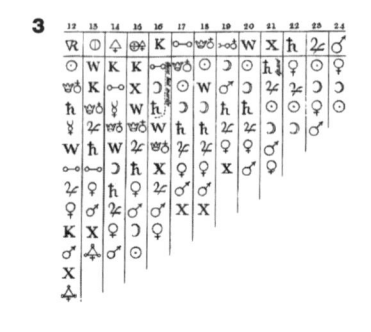

3
Alchemical chart of affinities in *Encyclopédie*, Diderot and D'Alembert, 1778.

4
Dmitri Mendeleev's periodic table II, 1871.

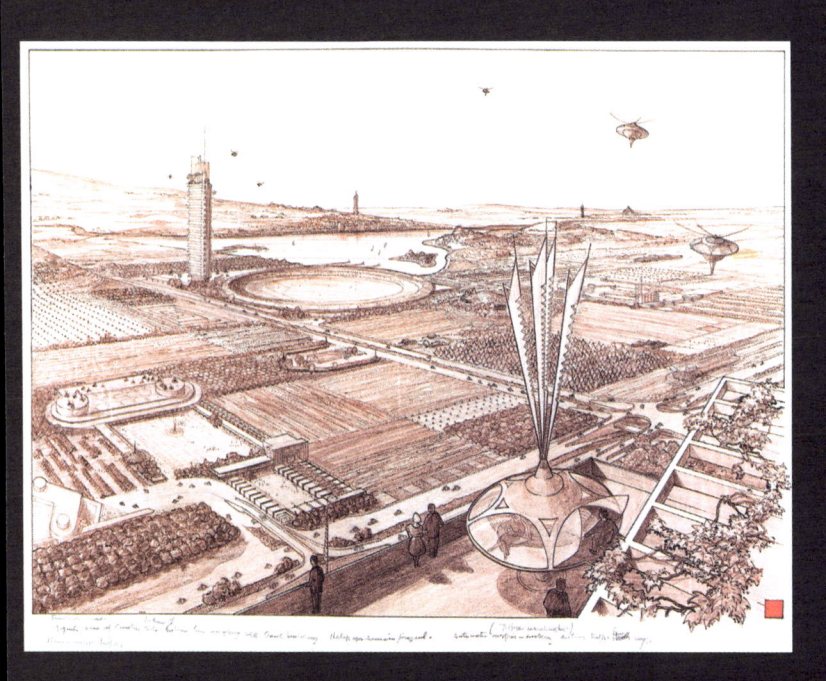

DISAPPEARING CITY?
Broadacre City, Frank Lloyd Wright, 1934–1958.

SÉBASTIEN MAROT

1
Agriculture/Industry,
François Kollar, 1931.

2, 3
Carl Bosch and Fritz Haber,
German chemists and
Nobel Prize winners.

4
World War I chemical warfare:
Poison gas attack on the western
front, 1915.

FROM WAR TO FIELD OPERATIONS

AGROCHEMISTRY AND MECHANIZATION

Liebig's discovery of the major role of nitrogen in plant growth did not immediately change agriculture. With no mineral source of nitrates, yields did not increase dramatically. The only known source was the guano gathered off the Pacific coast in Latin-America, mainly Peru: stocks of sea-bird faeces, preserved by the dry climate. At the end of the 19th century, Europeans were haunted by the "Malthusian trap" or "Malthusian spectre", in reference to the warnings of Thomas Robert Malthus (1766–1834) on the discrepancy he discerned between the geometric growth of population and the linear growth of food production or natural resources.

Nitrate shortage looked all the more absurd because of the abundance of atmospheric nitrogen. Hence, the synthetic capture of nitrogen became the new quest of chemistry. Fritz Haber (1868–1934) and Carl Bosch (1874–1940) succeeded in 1909–1910 in designing and industrialising a process for catalysing ammonia (NH_3). However, the process was extremely energy intensive and, therefore, very expensive, which made it hardly affordable for agriculture. But nitrates, derived from ammonia, are also a basis for explosives. During WWI, threatened by a shortage of explosives due to the blockade, the German Empire, soon emulated by the other warring countries, rushed to mass produce synthetic ammonia. At the end of the war, the European continent was thus equipped with a giant industrial infrastructure which would readily be converted into the production of fertilizers.

Indeed, WWI not only fostered the production of ammonia. It was also the cradle of other innovations that dramatically influenced the subsequent development of agriculture. Firstly pesticides were developed which were derived from the gas (organochlorine) produced for chemical war, another of Fritz Haber's feats. Secondly the first tractors were derived from tanks. Thirdly the war led to the massive production of barbed wire which would replace planted hedges. To this set of technologies WWII would of course add its own innovations, such as herbicides, originally developed to destroy the rice-fields of Japan. Even the origins of the modern food-processing industry could be traced back to the urgent need to feed millions of soldiers.

In other words, war economy not only accelerated but radically changed the course of agriculture and agronomy. The industrialisation of war was the matrix of the industrialisation of food and farming. Synthetic nitrates, pesticides and tractors became the pivots of a new agronomic system. War also prompted a new means of organising production that promised growth in productivity for both land and labour. This perspective exorcised, for a few decades, the *Malthusian specter*.

References
Max F. Peruz, "Le cabinet du docteur Fritz Haber",
La Recherche, no. 297, April 1997.
Sabine Jansen, "Histoire d'un transfert de technologie, De l'étude des insectes à la mise au point du Zyklon B", *La Recherche,* no. 340, March 2001.
A. Rougée, "L'histoire secrète des herbicides",
La Recherche, hors série no. 7, May 2002.

5
Fritz Haber instructing soldiers about chlorine gas deployment.

6
Chemical plant producing anhydrous ammonia.
Phot. Robert W. Kelley, 1954

URBAN RENEWAL
Jean-François Batellier, 1971.

1
Earl Butz, Head of the US
Department of agriculture
during the Nixon Administration,
Charles Bennett, 1975.

2
"A 21st Century Farm",
Davis Meltzer, 1970.

,3
Lettuce being sprayed in
California, Georges Steinmetz.
Date unknown.

4
"Mar de Plastico": Plastic
greenhouses in Spain, Georges
Steinmetz. Date unknown.

"GET BIG OR GET OUT"

INDUSTRIALIZING AGRICULTURE:
THE GREEN REVOLUTION

In the 19th century craft was industrialised. In the 20th, it was the turn of society as a whole, including agriculture and food production. Agriculture's traditional patterns were not appropriate to the new potential created by chemistry and mechanisation. Both land (small plots surrounded by hedges), species (mainly selected to brave rough conditions) and farmers (numerous, and wary of the effects and cost of these innovations) did not fit with the potential of the new products of industry. Industrialised agriculture required sources of huge amounts of capital, large enough to sustain a high level of investment (often obtained through debt), maximize the use of machines and make them profitable. Governments resolved to adapt landscapes, farmers, plants and animals to machines and fertilizers.

This so-called *"Green revolution"* — i.e. the industrialization of agriculture — was basically the result of strong public policies. Firstly, it required the furnishing of credit to encourage farmers to buy the machines. Secondly it required, the genetic selection of plants able to use high level of fertilizers and water. Thirdly land was consolidated to create larger plots and bigger farms, and finally fertilizers, pesticides and industrial seeds were subsidised. To that end, public authorities developed powerful agronomic research facilities. Banks were established that were devoted to agriculture. Systems of farm subsidies were created. In addition, the whole food supply chain was reframed, promoting actors engaged in industrialisation such as the farmers of middle-sized farms rather than small-holders or conservative landlords.

Industrialisation was a huge success in terms of food production. The yields per hectare were raised by a factor of two or three in a few decades. Labour productivity increased even more substantially: well-equipped workers in industrial countries could produce 120 times more than peasant farmers, working by hand. This led to a dramatic drop in the number of farmers (from 20 or 30% to less than 2 % of the working population in industrial countries today) and to a structural problem of overproduction which brought prices down. As the rural exodus accelerated, the food chain became longer and more complex. Food processing and retail sectors blossomed. Farmers often created cooperatives in order to keep or gain some power in the value-chain. Due to the high level of capitalisation, it became more and more difficult to keep farms within the family and pass them on to the next generation. As had been the case with craft production before the 19th century, the industrialisation of agriculture led to its *corporatisation*. Both motorisation and chemistry require large amounts of energy, mainly from fossil fuels. As a result, and unlike ecosystems which store solar energy in both ground and the biota, agriculture increasingly became a net energy consumer.

References
Henri Mendras, *La fin des paysans*, Paris, SEDEIS, 1967.
J. D. van der Ploeg, *The New Peasantries Rural Development in Times of Globalization*, Routledge, 2008.
Charles Mann, *The Wizard and the Prophet: Science and the Future of Our Planet*, Knopf, 2018.

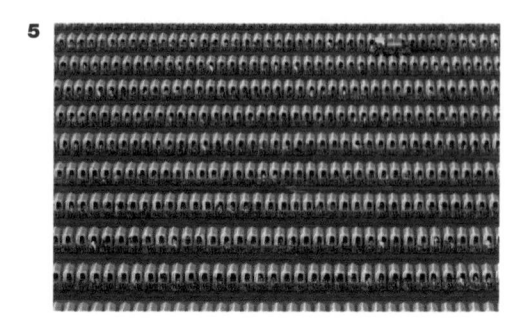

5
3,300 hutches for shelter newborn calves to supply the mega-dairies of Milk Source, George Steinmetz. Date unknown.

MINING FORESTS
Stan Eales, 1991.

SÉBASTIEN MAROT

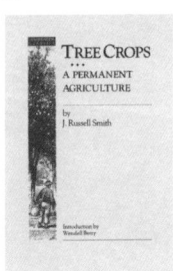

1
Farm village with piles of compost, Shantung, China, from *Farmers of Forty Centuries*, F. H. King, 1911.

2
View across Valley of Rice Fields, Kiangsi Province, China, from F. H. King, *ibid.*, 1911.

3
View of Fukuoka experimental station, Japan, from F. H. King, *ibid.*, 1911.

4, 5, 6, 7
Four major treatises on permanent and organic agriculture: 1911, 1929, 1940 and 1946.

PERMANENT AND ORGANIC AGRICULTURE

ALBERT HOWARD'S AGRICULTURAL TESTAMENT

Agriculture itself is not natural and in the past, by destroying soils and disrupting water cycles, it was often the main cause of desertification. But the disturbance caused by farmers, proportionate to their power, occurred slowly and often went unnoticed. The new powers procured by the industrial system were out of comparison with all that existed before, and indeed the environmental impact of industrial agriculture, be it in terms of biodiversity collapse, water use and pollution, or soil erosion, has been enormous. This, however, was highly predictable. Already in the first decades of the 20th century, insightful agronomists had indicated different pathways toward a *permanent agriculture*. While Franklin Hiram King (1848–1911) had documented the systematic composting and human waste recycling in the East, J. Russell Smith (1874–1966) famously pleaded for a "two-story agriculture" based on perennial forest ecosystems. However at the onset of the industrialisation of agriculture, the most powerful lessons and warnings came from English agronomist Albert Howard (1873–1947).

A son of farmers, Howard studied agriculture at Cambridge, but spent most of his career in tropical countries, especially India. He admired the skills of traditional farmers in the tropics, and praised their ability to produce high yields while maintaining land fertility. He concluded that the keystone of sustainability was the careful recycling and management of organic matter, and claimed that industrial agriculture would poison the environment; waste organic matter, lead to a rapid soil erosion and finally to a drop in productivity. As an alternative, he advised intensifying production not through chemistry but by mimicking ecosystems and particularly the forest which he considered the most efficient one. In this, he put himself at odds with most of his colleagues and the scientific establishment. Back to England, he summarised his arguments in two major books: *An Agricultural Testament*, and *Soil and Health*. In 1946, he supported Lady Evelyn Barbara Balfour (1898–1990) and her Soil Association, which played a prominent role in the early development of *organic farming*.

Public concern was later accelerated by Rachel Carson's famous *Silent Spring* (1962), which denounced the effects of pesticides on wildlife (especially bird populations), and provoked a strong counter reaction from the agroindustry, eager to deny its long-lasting impacts on both human health and biodiversity. Today, the dire consequences of industrial agriculture are less and less contested. With climate change adding a whole new level of complexity and threat, the Food and Agriculture Organization has recognised the need to deeply rethink the trajectory of agriculture. But the system's lock-in is so powerful that despite the growing awareness of consumers and some evolution in both law and consumption patterns, the train of industrialisation keeps going down the same track.

References
F. H. King, *Farmers of Forty Centuries, or Permanent Agriculture in China, Korea and Japan,* Madison, 1911.
J. Russell Smith, *Tree Crops: A Permanent Agriculture*, Harcourt, Brace and Co, 1929.
Albert Howard, *An Agricultural Testament*, Oxford Univ. Press, 1943.
Albert Howard, *The Soil and Health: A Study of Organic Agriculture*, Faber & Faber, 1947.
Fairfield Osborn, *Our Plundered Planet*, Boston, Little, Brown and Co, 1948.
Rachel Carson, *Silent Spring*, Houghton Mifflin, 1962.

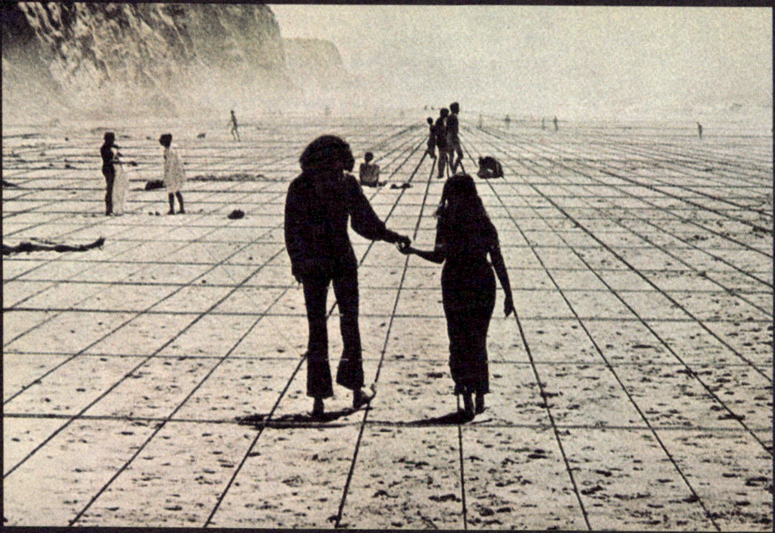

1
Green Business.

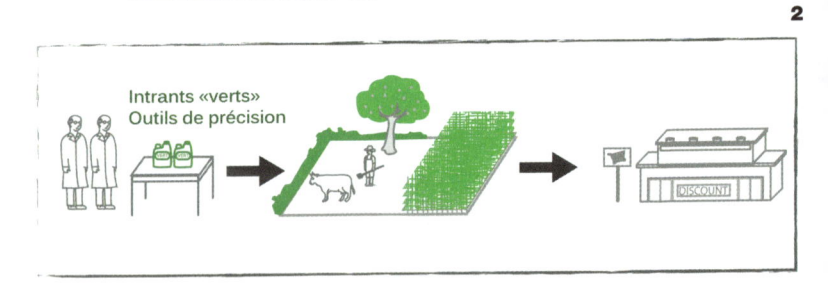

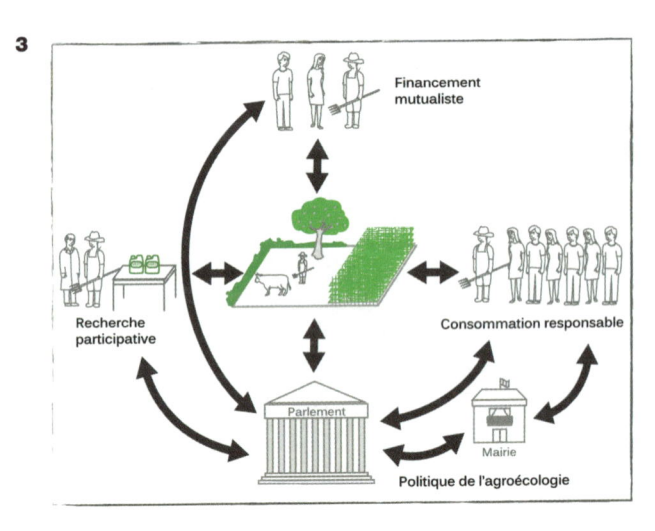

2
Soft Agroecology.

3
Strong Agroecology.

BEYOND INDUSTRIAL AGRICULTURE

CYBORGS VS AGROECOLOGY

Nowadays, there is little argument that the industrial model of "the green revolution" is not sustainable. In the last few years, many new concepts have flourished: conservation agriculture, climate smart agriculture, integrated pest management, ecologically intensive agriculture, etc. In parallel, the concept of agroecology, coined in the 1920s to name a new approach to agronomy based on the lessons of dynamic ecology, has resurfaced and gained momentum as a broader discipline that embraces the social and political aspects of the agrarian question. If all these concepts and strategies pretend to aim at the same result — reconciling agriculture and the environment — they deeply differ in scope, and on the means and actors they respectively involve.

High tech agriculture and green business are focused on big data, technical innovations and market, and promote the agriculture of a new industrial revolution. They are all about changing the farmers' tools, or feeding farms with "greener" inputs, but without questioning or challenging the system itself.

Soft agroecology acknowledges the need to redesign balanced agrosystems that would mimic or emulate ecosystems (biomimicry), but without changing the value-chain. Agriculture remains a business of professionals.

Strong agroecology emphasizes the need to redesign balanced agrosystems, but also that of rethinking the whole food system from labs to kitchens. The quest for sustainability in agriculture is inextricably a social question that needs to be politically addressed.

Spiritual agroecology holds that sustainability is well beyond the reach of mere rationality and requires a dramatic reassessment of our relation to Nature. The metamorphosis needed is so radical that only a new awakening or religio mundi, by harnessing the forces of our emotions and feelings towards ourselves, our biotic communities and our surroundings, might put it into gear.

Collapsologues argue that jurisprudence on the numerous collapses of former complex societies, as well as our obvious present incapacity to change course despite all scientific warnings, tend to prove that, at some point, civilizations are more likely to disappear rather than change. As to which new societies will emerge from there and what they might look like… no one knows.

References
Miguel Altieri, *Agroecology : the Science of Sustainable Agriculture*, Westview Press, 1995.
Pablo Servigne, *Nourrir l'Europe en temps de crise : vers des systèmes alimentaires résilients,* Nature et Progrès, 2014.
Matthieu Calame, *Comprendre l'Agroécologie: origines, principes et politiques*, Charles Léopold Meyer, 2016.

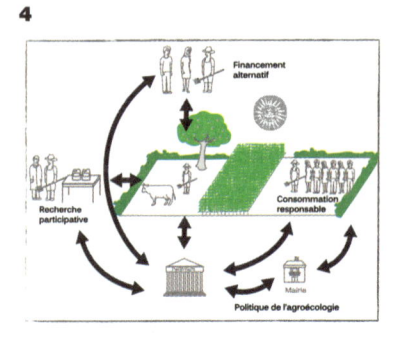

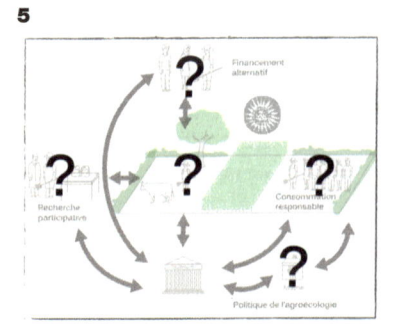

4
Spiritual Agroecology.

5
Collapsology.
Comprendre l'Agroécologie,
Matthieu Calame, 2016.

INHABITING THE WOODLAND
Notre Dame des Landes ou le métier de vivre,
Christophe Laurens and DSAA Alt-U Students, 2018.

D.

EXIT URBS

In which the puzzled architect realises that maybe cities and metropolises aren't the manifest destiny of humankind they pretend to be, that a long tradition of counter-prophets did indeed embrace the country's side (in the name of agrarianism, ruralism or Back-to-the-Land), and that their heirs might have good insights to share as to the burning question: "where do we go from here?"

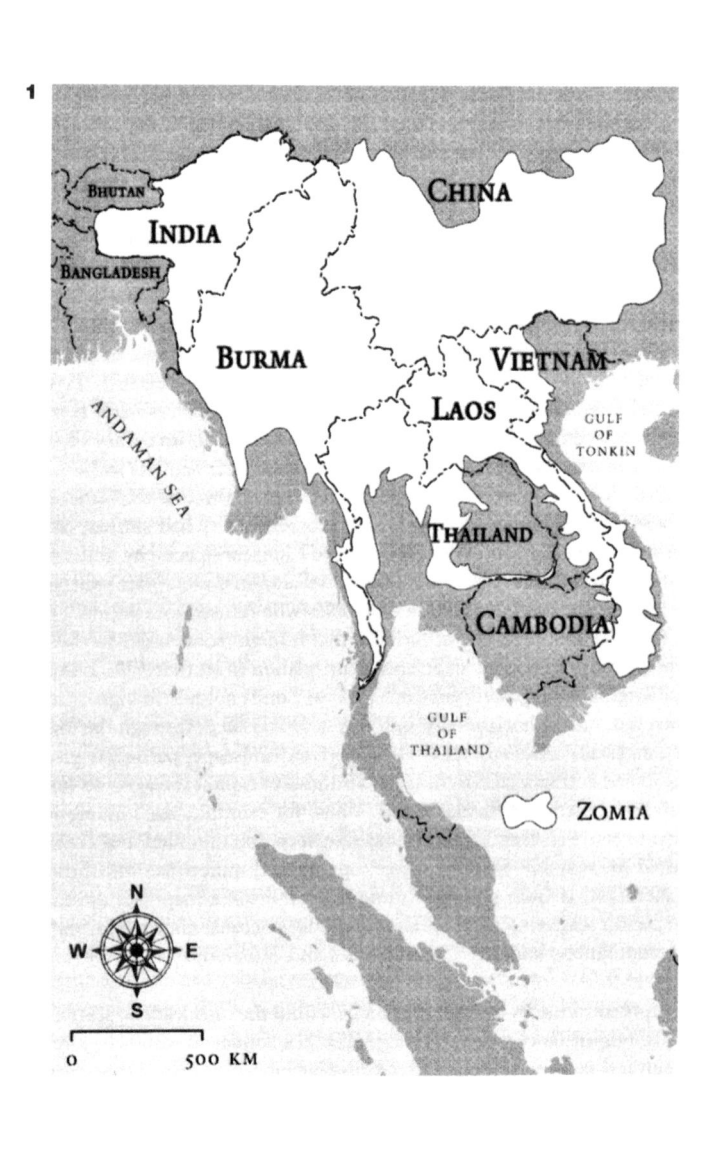

1
**Zomia: Uncontrolled territories
in South-East Asia Highlands,
James C. Scott, 2009.**

AGRICULTURE AND ARCHITECTURE

ZOMIA

BACK-TO-THE-LAND?
OR FORWARD INTO
THE HILLS AND FORESTS?

Taking up the provocative thesis put forward in the 1960s and 1970s by anthropologist Pierre Clastres, agrarianist and historian James C. Scott has famously proposed a counter-history of "barbarians". In his view, most of the "savage" or raw tribes described by early chronicles, and later studied by anthropologists, were not so much primitive societies yet to be civilized than a genuine by- or counter-product of early cities and states: i.e. "barbarians by design", eager to evade the conditions of serfdom and domination imposed on their populations by city-states, with their complex societies and political structures. Hence, Back-to-the-Landers could be as ancient as cities themselves. They even probably played a significant role in the erosion and collapse of states and urban infrastructures, at least they did until the modern era when virtually every recess of our planet was subject to political enclosures.

In *The Art of Not Being Governed* (2009), Scott looked more specifically at the history of upland Southeast Asia, a vast region (almost the size of Western Europe) comprising Northeast India, Burma, Southeast China, and extending far south into Thailand, Laos, Vietnam and Cambodia: a kind of "giant Switzerland without cuckoo clocks" which he designates as Zomia. Scott shows that it was relentlessly peopled by fugitives who tried to escape the expanding orbit of states that took hold of government in the adjacent valleys and lowlands. In contrast to the simplified and labor intensive agrosystems developed by those states, based on the concentration of irrigated annual and easily taxable crops (rice paddies) — with their tolls of pests and diseases — these barbarians resorted to (and evolved) a whole range of varied site specific agricultural techniques (and social structures) specifically tailored so as to avoid state control and conscription, thereby securing their autonomy.

In his book, Scott insists that most of the millenarian prophets who radically questioned the sustainability of large state societies and their values (rather than simply plead for the redistribution of their roles or the reorganisation of their plumbing) came from the uplands or hinterlands. There, other kinds of contracts with the environment, and another intelligence of nature, were experimentally entered into. Indeed, those who, today, consciously take the country's side, might well be the most significant heirs of these people.

References
Pierre Clastres, *La Société contre l'État : Recherches d'anthropologie politique*, Minuit, 1974.
James C. Scott, *The Art of Not Being Governed: An Anarchist History of Upland Southeast Asia*, Yale, 2009.

2, 3, 4
Ethnic minority communities of the Vietnamese Highlands, Sascha Richter, 2016.

MILITARY INDUSTRIAL ECOLOGY?
A Marine HRS-1 Helicopter flies away with its own hangar,
Buckminster Fuller, 1954.

SÉBASTIEN MAROT

1
Monte Cassino, seat of the Benedictine Order, Italy, established around 529. Francesco Cassiano De Silva, 1703

2, 3
Forest Cistercian retreats, Thoronet Abbey, Var, France, 12th century and Senanque Abbey, Gordes, Luberon, France, founded in 1148.

AGRICULTURE AND ARCHITECTURE

MONASTERIES

TAKING THE COUNTRY'S SIDE

The history of civilisation is often presented as an irrepressible march forward to ever larger and more complex forms of organisations. However, one should not forget that it was also a long chronicle of collapses, often much more profound than the surge of brilliancy they followed and eroded. As these moments leave less obvious traces in landscapes and official records, they have tended to be overlooked (if not frankly disregarded) by modern historians. Such a period began in Europe with the fall of the Western Roman Empire around the 4th and 5th century AD. It unleashed a massive dispersion and ruralisation into smaller communities, a disappearance or shrinkage of cities for which early historians (between the Renaissance and the 17th century) retrospectively coined (in a surge of militant shortsightedness) the terms of Middle or Dark Ages. A renewed understanding and reappraisal of that fascinating period of European history is on its way, and badly necessary.

Useful lessons in communal politics and local resource stewardship could be drawn from a renewed look at *encellullement* in villages and counties that took place across Europe during those centuries: a phenomenon of spatial and political contraction into smaller vernacular communities. In the light of what the urban condition has become nowadays, no doubt they would rapidly challenge the assumption that only "Stadtluft macht frei", and cast that prejudice as a typical bias of modern obscurantism.

Religious orders and monasteries (Benedictine, Cistercians, etc., but also pilgrims) were at the forefront and vanguard of this movement of exile and recollection in remote rural and forested areas, where sisters and brothers invested their energy and care in both the cultivation of the earth (developing a whole range of agricultural practices and techniques) and the salvaging, treasuring, sifting and transmission of a cultural legacy. In their bareness and austerity, Romanesque monasteries in particular were exemplary of this *encellulement*. Situated in distant deserts and forests, they turned them into teeming mundi: inland islands and spiritual airports within the larger world they kept at bay. Indeed, the lean perfection they jointly attained in architecture, agriculture and craftmanship has hardly been surpassed in the West ever since.

References
Jérôme Baschet, *La civilisation féodale: de l'an mil à la colonisation de l'Amérique*, Flammarion, 2006.
Joelle Zask, *La Démocratie aux champs, du jardin d'Éden aux jardins partagés, comment l'agriculture cultive les valeurs démocratiques*, La Découverte, 2016.

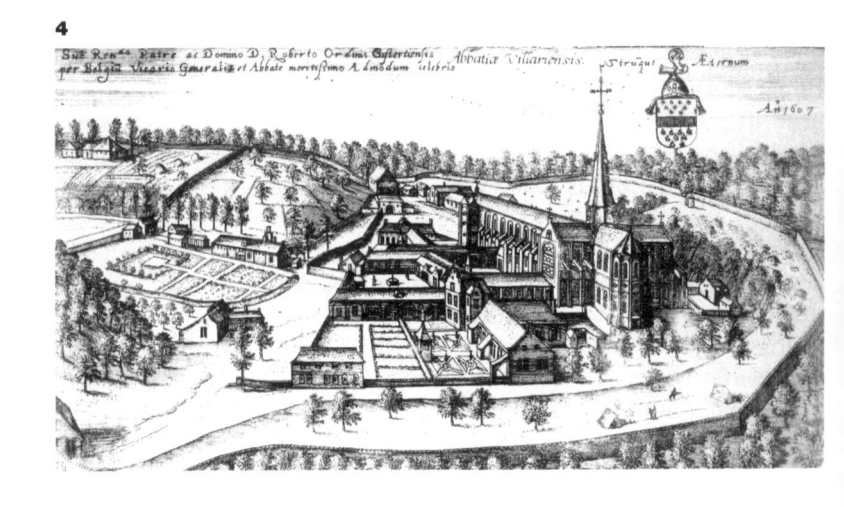

4
Inland Island, Villers Cistercian Abbey, Belgium, founded in 1146, 1607.

84 What can be done with freeways after the automobile age?

PROSPECTIVISM
Wasting Away, Kevin Lynch, 1990.

SÉBASTIEN MAROT

1

2 PETER KROPOTKIN — FIELDS, FACTORIES and WORKSHOPS — Edited by George Woodcock

3 LIGHT FROM THE CITY — RALPH BORSODI — ABUNDANT SECURITY ON THE LAND

4 THE LIMITS OF THE CITY — Murray Bookchin

5 MURRAY BOOKCHIN — toward an ecological society

ANARCHY AND LOCALISM

TOWARDS HOMESTEADS
AND COMMUNITIES' SELF RELIANCE

Often misrepresented as an irresponsible advocation of chaos, anarchism is a tradition of thinking and practical experiments which has in fact contributed a significant body of reflection on the conditions and scales at which responsibility, relative independence and resilience may be achieved (or safeguarded) at both the individual and collective levels. Its links to agrarianism are quite strong, and, arguably, no other political line of thinking has a comparable record of contributions to the development of ecology and environmental reflection over the past two centuries.

Among those anarchist thinkers who exerted a deep influence on urbanists and architects such as Patrick Geddes, Lewis Mumford or Frank Lloyd Wright, Russian geographer and naturalist Peter Kropotkin (1842–1921) stands out as a particularly important figure. Aside from his work as a zoologist, and his contribution to Darwinian Theory, where he emphasized the role of "mutual aid" or co-operation as a "factor of evolution", Kropotkin famously pleaded, in *Fields, Factories and Workshop* (1898), for a local integration of agriculture, industry and craftmanship into largely self-reliant and self-governing small communities.

A few decades later, during the American Dustbowl, another powerful surge of local agrarianism came with Ralph Borsodi's book *Flight from the City: An Experiment in Creative Living on the Land* (1933) which advocated, against centralised systems where "distribution costs tend to move in inverse proportion to production costs", the reuniting of home and workplace in modular communities where each household would become a local production centre providing most of its basic resources. Finally, in the post-war period, it was plausible for anarchists to make a new plea for the end of cities and the rise of communities based on an indictment of the deleterious environmental consequences of massive urbanisation, An original (and somewhat overlooked) figure in that respect was architect Erwin Anton Gutkind (1886–1968) who coined the concept of *social ecology*. But the key thinker in that vein (influenced by both Mumford and Gutkind) is certainly left activist Murray Bookchin (1921–2006) who relentlessly coupled his early indictment of capitalism as the main cause of environmental degradation with a plea for the building of eco-communities.

References
Peter Kropotkin, *Fields, Factories and Workshop*, Hutchinson, 1898.
Peter Kropotkin, *Mutual Aid: A Factor of Evolution*, McClure, Phillips & Co, 1902.
Ralph Borsodi, *Flight from the City: An Experiment in Creative Living on the Land*, Harper and Brothers, 1933.
Erwin Anton Gutkind, *Community and environment. A discourse on social ecology,* Watts and Co, 1953.
Erwin Anton Gutkind, *The Expanding Environment. The end of cities, the rise of communities*, Freedom Press, 1953.
Murray Bookchin, *Our Synthetic Environment*, Knopf, 1962.
Murray Bookchin, *The Limits of the City*, Harper, 1973.
Murray Bookchin, *The Rise of Urbanization and the Decline of Citizenship,* Sierra Club, 1987.

1
A self-sufficiency sight: Three Acre Homestead, *Flight from the city*, Ralph Borsodi, 1933.

2, 3, 4, 5
Toward decentralization and political ecology, 1891, 1933, 1973, 1980.

FEEDING THE PETRI DISH
Architecture of density, Michael Wolf, 2014.

1
Liberty Hyde Bailey plowing the first furrow of the new State College of Agriculture at Cornell, 1905.

2
Champion of the rural world: 32nd Annual Cornell Dinner menu, March 5, 1912.

RURALISM

A PRACTICAL PHILOSOPHY OF COUNTRY-LIFE

Liberty Hyde Bailey (1858–1954) was raised on the Michigan Frontier before becoming a renowned botanist and horticulturist, not to mention the founder of the New York State College of Agriculture at Cornell University. He was a lifelong advocate of the cause of farmers, the condition of the rural world, and the conservation of the "backgrounds" of society.

In Bailey's view, agriculture was not just a congeries of crafts and applied sciences but fundamentally a culture, a stewardship of land, soil and nature, an ethic of life and resilience which it was the mission of the State College to develop and sustain in all its aspects and dimensions. The rural exodus caused by the industrial revolution, and more generally the progressive drainage of human and material resources from the country towards the cities and metropolises, had to be checked at source (instead of being merely accommodated in their consequences). In other words, a whole philosophy was needed, for which he coined the term "ruralism", to complement (and relativise) the nascent discipline of urbanism.

This agenda, which consisted in celebrating and improving the desirability of agriculture and rural life, Bailey pursued it not only at Cornell, by envisioning the whole campus as an illustration and laboratory of this ruralism, and by launching a department of "rural art" — i.e. landscape design — in the College. He was also appointed by Theodore Roosevelt as president of the federal Commission on Country Life . Given his all-encompassing views, many were disappointed that he eventually declined President Taft's offer to be his US Secretary of Agriculture, because of his scientific and academic commitments and a certain suspicion about politics. But Bailey, who had been all along an active organizer of the Nature Study Movement (which strived to acquaint school kids with nature and gardening), did instead distill, in a series of amazing books, a practical philosophy and ethic of the land, whose legacy is steadily being rediscovered and mined by contemporary environmentalists.

References
Liberty Hyde Bailey, *The State and the Farmer*, Mac Millan, 1908.
Ben Minteer, *The Landscape of Reform: Civic Pragmatism and Environmental Thought in America*, MIT, 2006.

3
Nature study: Ithaca's public
school, children garden, 1907.

NEW ALCHEMY
New Alchemy Institute, Nancy & John Todd, 1969.

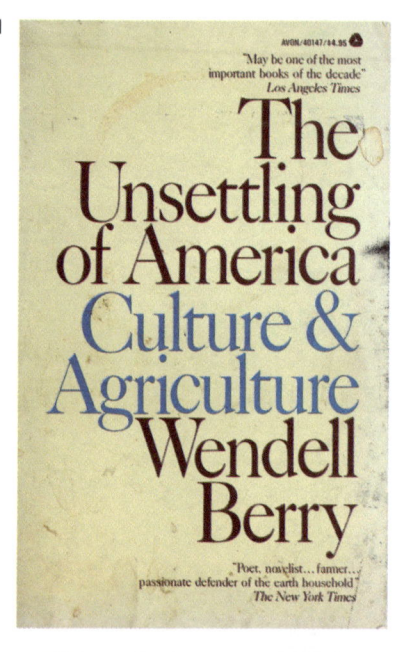

1
A plea for localism against agribusiness, 1977.

2
Wendell Berry working on family farm in Port Royal, Kentucky, *Look & See: A Portrait of Wendell Berry*, 2016.

AGRICULTURE AND ARCHITECTURE

"THINK LITTLE"

A PROSECUTION OF AGRI-BUSINESS

Among the few beacons of agrarianism that have emerged since World War II, novelist, poet, essayist and farmer Wendell Berry (b. 1934) is certainly one of the most abiding and consistent. An early advocate of organic farming and agro-ecology, and a fierce environmental activist, Berry has been above all a eulogist of the local, of place connectedness, a champion of small community cultures and traditions, site-specificity and "appropriate technology", all concerns which led him, for instance, to praise the way of living of Amish communities.

In "Think Little", a short essay originally published in the *Whole Earth Catalog in* September 1970, Berry pointedly argued that the environmental movement could not satisfy itself by just making public complaints and demands for big changes (in laws, policies and politics). It was doomed if it did not advance individual consciousness and behavior, rooted into the active commitment to put ones house or backyard in order: "If you are fearful of the destruction of the environment, he wrote, then learn to quit being an environmental parasite... Odd as I am sure it will appear to some, I can think of no better form of personal involvement in the cure of the environment than that of gardening. A person who is growing a garden, if he is growing it organically, is improving a piece of the world... and less dependent on an automobile or a merchant for his pleasure."

In 1977, with the publication of *The Unsettling of America: Culture and Agriculture*, Berry's powerful coupling of agrarianism and environmentalism was fully articulated as a plea for small-scale organic farming and an indictment of the "get big or get out" logic of agri-business: "Because the soil is alive, various, intricate, and because its processes yield more readily to imitation than to analysis, more readily to care than to coercion, agriculture can never be an exact science. There is an inescapable kinship between farming and art, for farming depends as much on character, devotion, imagination, and the sense of structure as on knowledge. *It is a practical art,*" wrote Berry. As such, farming does not require doctrine, but skill:

"Skill... is the enactment or the acknowledgment or the signature of responsibility to other lives; it is the practical understanding of value. Its opposite is not merely unskillfulness, but ignorance of sources, dependencies and relationships."

References
Wendell Berry, "Think Little", in *Whole Earth Catalog*, Sept 1970.
Wendell Berry, *The Unsettling of America: Culture and Agriculture*, Sierra Club, 1977.

3
Wendell Berry at his desk in the 1970s, Kentucky.

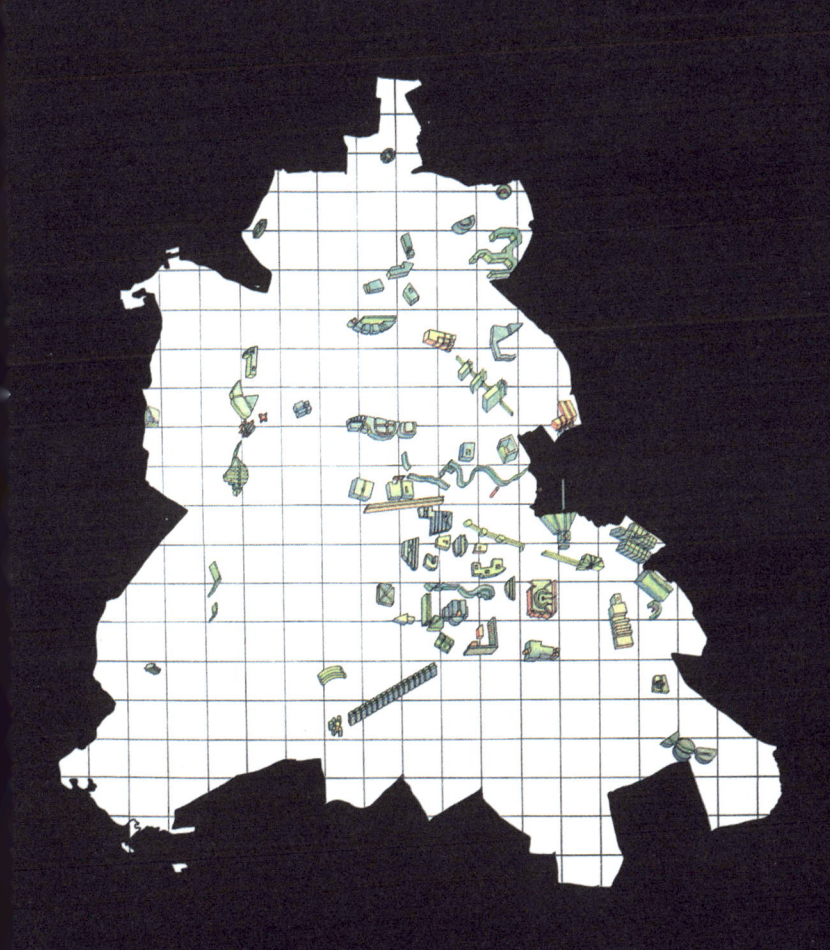

ARCHIPELAGO
The City in the City — Berlin: A Green Archipelago,
Oswald Mathias Ungers & Rem Koolhaas, 1977.

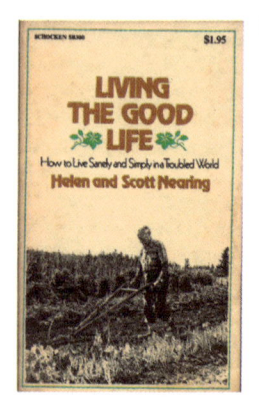

SÉBASTIEN MAROT

1
The grandparents of the Back to the Land Movement: 1973 (1st edition 1954).

2
The Last Whole Earth Catalogue, Stewart Brand, 1971.

3
Supplement to the Whole Earth Catalog, 1971.

"WORKERS OF THE WORLD, DISPERSE"

BACK-TO-THE-LAND AND COUNTERCULTURE

In the wake of the Great Depression, but also in the aftermath of World War II and during the three decades of growth that followed it, several intellectuals, pacifists and veterans advocated and embraced a kind of voluntary exile from the consumer society and from what would later be stigmatized as the "military-industrial complex". In the US, Scott and Helen Nearing's *Living the Good Life* (1954) was particularly influential. In the book, these famous socialists recounted their experience in self-sufficient domestic farming in Vermont and Maine. When it was republished in 1970 its authors were branded as the grand-parents of the back-to-the-land movement.

Indeed, it was only then, toward the end of the 1960s, in a climate marked by the Vietnam War, the hippies and the moon shot that the said movement really gained momentum and registered in the statistics. One of its main organs in the US, which was emulated in other countries (i.e. the *Catalogue des Ressources* in France), was the *Whole Earth Catalog*, a remarkable publication, launched by biologist Stewart Brand, which promoted all kinds of intellectual resources, practical tools and alternative technologies (in fields as diverse as gardening, masonry and communications) for those who were eager to flee the "system" in order to experience other and more sustainable ways of living.

A wide-ranging collector of major threads of alternative thinking (Lewis Mumford, Buckminster Fuller, Ernst Schumacher, Wallace Stegner, Wendell Berry, etc.) and contributions from key figures and writers of the counterculture (Ken Kesey, Robert Crumb, Gary Snyder, etc.), the *Whole Earth Catalog* was, during its initial 4 years (1968–1972) an essential catalyst for do-it-yourself whose legacy is claimed by both contemporary environmentalism and cyberculture.

As a footnote, it may be worth mentioning that Stewart Brand drastically has changed his position since. He now advocates *concentration* (dense cities, nuclear power, GMO and geoengineering) as the only "rational" solution to the threats posed by climate change and other ecological concerns... and brandishes this stunning about-turn as a certificate of both intellectual courage and environmental pragmatism.

References
Helen and Scott Nearing, *Living the Good life: How to Live Sanely and Simply in a Troubled World*, 1970.

4
Geodesic thoughts, *The Last Whole Earth Catalogue*, 1971.

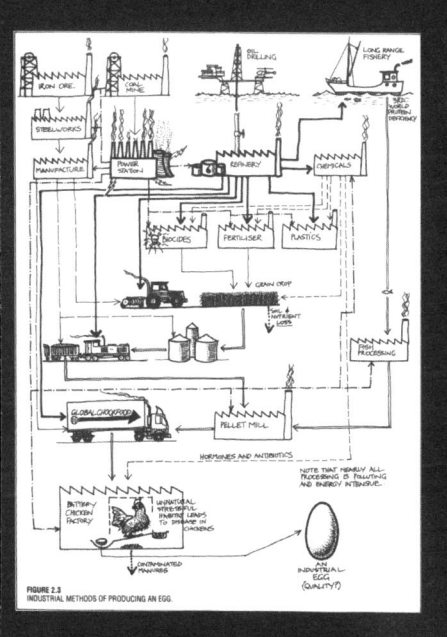

FIGURE 2.3
INDUSTRIAL METHODS OF PRODUCING AN EGG.

FIGURE 2.4
PERMACULTURE METHODS OF PRODUCING AN EGG.

SHORT SUPPLY CHAIN
Permaculture, a designers' manual, Bill Mollison, 1988.

1. *New Roots for Agriculture*, Wes Jackson (Friends of the Earth)

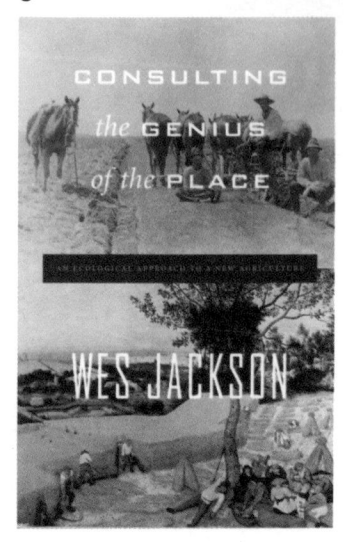

2. *Becoming Native to This Place*, Wes Jackson

3. *Consulting the Genius of the Place*, Wes Jackson

AGRICULTURE AND ARCHITECTURE

BECOMING NATIVE

LAND INSTITUTE: THE QUEST
OF "NEW ROOTS FOR AGRICULTURE"

Raised on a farm in Kansas, Wes Jackson (b. 1936) studied biology, botany and genetics before serving as chair of one of the US's first environmental studies program at California State University. In the mid-1970s, he left academia and moved back to Kansas where, together with his wife Dana, he founded the Land Institute, a non-profit organisation dedicated to developing a natural systems agriculture that mimic or emulate the way local ecosystems naturally function.

For more than four decades, the Land Institute has conducted experiments into perennial polycultures, i.e. associations of perennial grains, pulses and oilseeds, which, because they are both self-sustainable (requiring no chemical fertilisers or pesticides) and soil structuring, could advantageously replace the giant monocultures of annuals that have eaten up the American prairie and considerably eroded and degraded its soils.

Alongside his work at the Institute, Jackson has published a series of major books and essays which expose the need for rethinking agriculture and its place in economy and society. Central to his philosophy is the idea that any environmental programme that does not gives precedence to agriculture is futile and doomed to fail. He wrote: "It is my view now that if we don't get sustainability in agriculture first, it is not going to happen, and for this reason that agriculture alone, ultimately, has a discipline behind it. It is the synthetic discipline of ecology–evolutionary biology. Nature's ecosystems are ancient. They are real economies. The law of return described by Sir Albert Howard operates. They can be trusted. The materials sector, the industrial sector, is recent. It has no time-honored discipline to draw on. Skyscrapers, freeways, and suburbia have been made possible due to our discovery and use of fossil carbon, not any organising concept. Soils and forests have fed and sheltered us, but they too are in decline even with the subsidy of fossil fuels."

In the end the only viable agenda is relocalisation: "We are unlikely to achieve anything close to sustainability in any area unless we work for the broader goal of becoming native in the modern world, and that means *becoming native to our places* in a coherent community that is in turn embedded in the ecological realities of its surrounding landscape," he writes.

References
Wes Jackson, *New Roots for Agriculture*, Friends of the Earth, 1980.
Wes Jackson, *Becoming Native to This Place*, University Press of Kentucky, 1994.
Janine Benyus, *Biomimicry: Innovation Inspired by Nature*, William Morrow, 1997. First chapter: "How will we feed ourselves? Farming to fit the land: growing food like a prairie".
Wes Jackson, *Consulting the Genius of the Place: An Ecological Approach to New Agriculture*, Counterpoint, 2011.

1
A plea for perennial plants, *New Roots for Agriculture*, Wes Jackson, 1980.

2, 3
Rerooting, 1994, 2011.

4
Wes Jackson and E.F. Schumacher at the Land Institute, 1977.
A New Roots Special Report, 1982.
Phot. Terry Evans, 1977

VOCATIONAL RETRAINING
Déchets, Julie Nahon, 2019, from the painting
Des Glaneuses, Jean-François Millet, 1857.

E.
FACING THE PRESENT ENVIRONMENTAL PREDICAMENT

In which we realise, in bewilderment, the extent to which the environmental mess we are currently in was documented and predicted almost 50 years ago, and how these warnings inspired a trove of reflections, both profound and practical, on the implications for technology and design from outside the sphere of experts and professionals.

1

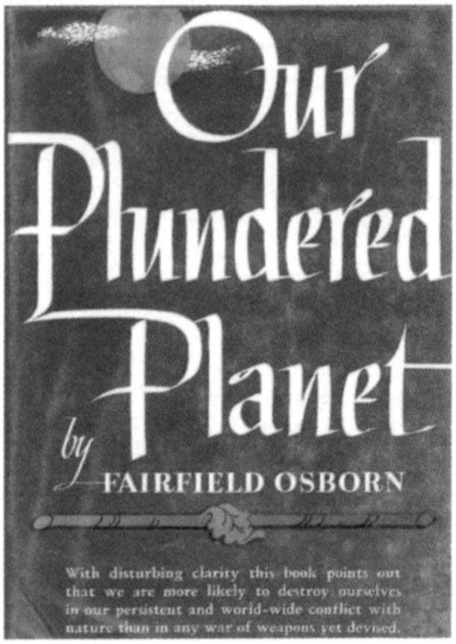
Our Plundered Planet
by FAIRFIELD OSBORN
With disturbing clarity this book points out that we are more likely to destroy ourselves in our persistent and world-wide conflict with nature than in any war of weapons yet devised.

2

A Sand County ALMANAC
AND SKETCHES HERE AND THERE
By ALDO LEOPOLD

3

50TH ANNIVERSARY EDITION
THE DEATH AND LIFE OF GREAT AMERICAN CITIES
JANE JACOBS
With a new Introduction by Jason Epstein

4

FIFTIETH ANNIVERSARY EDITION
'Carson's book has changed the world'
The Times
SILENT SPRING
Rachel Carson
WITH A NEW INTRODUCTION BY CAROLINE LUCAS

5

FIFTY CENTS
FEBRUARY 2, 1970
TIME
Environment: Nixon's New Issue
ECOLOGIST BARRY COMMONER
The Emerging Science of Survival

6

printemps 72
n° 11
SURVIVRE
... et Vivre

"A BLUEPRINT FOR SURVIVAL"

THE MOUNTING ALARM OF ENVIRONMENTALISM

Environmental concerns did not suddenly appear in the 1960s. Throughout the industrial era, several voices repeatedly pointed out ecological predicaments, detailing the blind spots and dead ends of industrial development, and the devastation of entire regions and colonies by capitalism. On the whole, these were basically silenced by the pervasive dogma of productivism, a meta-religion (as described by its high priests Saint-Simon, Comte, etc.) which was adhered to by a wide political spectrum going all the way from the champions of the free market to their fiercest opponents (i.e. the Marxist doxa, and its so-called "dialectics").

During the first half of the 20th century, as this modernist dogma became more and more dominant, major events and crises, such as the Great Depression, sparked the first critiques. Lewis Mumford, both as a historian in his books *Technics and Civilization* (1934) and *The Culture of Cities* (1938), and as a promoter of "regional planning" was one of the first to give these voice. In the wake of WWII and its nuclear ending, other thinkers, such as Jacques Ellul, deepened this critique of the technological system. In addition, a number of books, such as Aldo Leopold's *A Sand County Almanac* (1949) started to formulate a "Land Ethic" but again, Reconstruction fever morphed into the militant optimism and "addiction to growth" of Les Trentes Glorieuses — the three boom decades after WWII.

Indeed, it is only in the 1960s that these threads of critique started to coalesce into a conscious environmental movement. In the USA, two books by women authors and scientists, respectively devoted to the consequences of modernism in urbanism and agriculture, played a crucial role in this awakening: Jane Jacob's *The Death and Life of Great American Cities* (1961) and Rachel Carson's *Silent Spring* (1962), arguably the most influential book ever written in the field. In the following years, culminating in the early 1970s — particularly between Earth Day on April 22 1970 and the UN Stockholm Conference on the Environment in June 1972 — many scientists documented the interrelated aspects of the environmental predicament. Biologist Barry Commoner in his book *The Closing Circle: Nature, Man, and Technology* (1971), and the global "Blueprint for Survival", a key issue of the British journal The Ecologist is a good example. Others, such as mathematician Alexandre Grothendieck, questioned the very programs and responsibilities of scientific research and institutions. If most of these whistleblowers came from hard and life sciences, some — rarer, perhaps — launched their arguments from within the field of economics: in *The Entropy Law and The Economic Process* (1971), Nicholas Georgescu-Roegen, who was particularly well-versed in agrarian issues, demonstrated that no sound economic reasoning could ignore the second law of Thermodynamics.

References
Serge Audier, *L'Age productiviste, hégémonie prométhéenne, brèches et alternatives écologiques*, La Découverte, 2019.

7

1, 2
Postwar breakthroughs in environmental consciousness and ethics: 1948, 1949.

3, 4
Questioning the path of modern urbanism and agriculture: 1961, 1962.

5, 6
Science and survival: Barry Commoner 1970, Alexandre Grothendieck 1972.

7
Collective Manifesto: "A Blueprint for Survival", *The Ecologist,* **Edward Goldsmith, January 1972.**

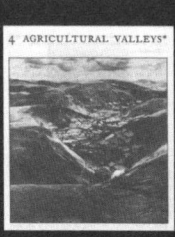

A GENERATIVE GRAMMAR FOR DESIGN
A Pattern Language: Towns — Buildings — Construction,
Christopher Alexander, Sara Ishikawa & Murray Silverstein, 1977.

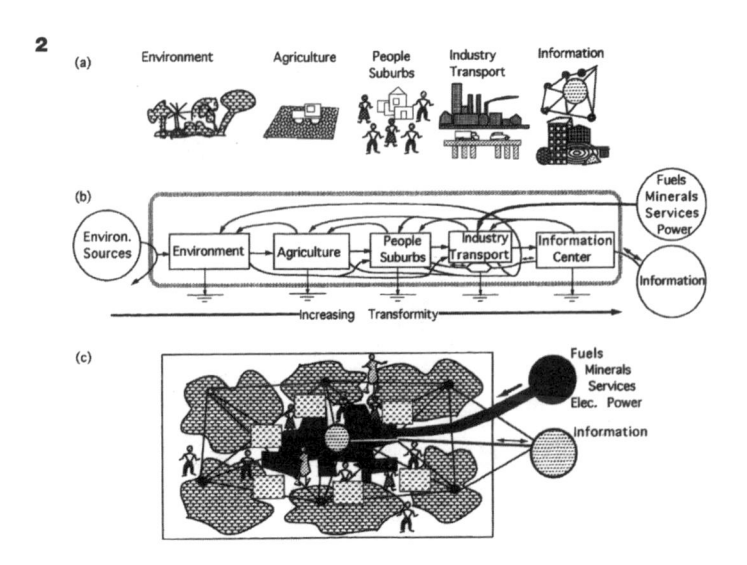

1
Model of the past agrarian landscape mostly running on renewable energies. Elisabeth and Howard Odum, *ibid.*

2
Model of the fuel-based landscape of urban America with suburban populations outside of the fuel-using transport-industrial areas. Elisabeth and Howard Odum, *ibid.*

AGRICULTURE AND ARCHITECTURE

ENERGY DESCENT

ECOLOGY
AND SYSTEM'S ENERGETICS

The concept of ecology was originally coined in 1866 by Ernst Haeckel and defined by him as the "comprehensive science of the relationship of the organism to the environment". In the following decades, it slowly emerged, alongside physiology and morphology, as one of the major subsections of biology, and was framed as the "philosophy of living nature" by John Burdon Sanderson in 1893. Still later, as it developed its own concepts, scales and hierarchies such as the ecosphere, biomes, ecosystems, etc., it also incorporated elements from mineral chemistry, energetics and system's dynamics. When it did so ecology was finally established as a genuine science and discipline.

After the pioneering work of botanist Arthur Tansley (1871–1955) and limnologists George Evelyn Hutchinson (1903–1991) and Raymond Lindeman (1915–1942), crucial in this respect were the contributions of Eugene P. Odum (1913–2002) and Howard T. Odum (1924–2002): sons of a renowned sociologist and regionalist thinker whose work had been closely related to Lewis Mumford's. The seminal and most influential textbook which the two brothers co-authored in 1953, *Fundamentals of Ecology*, could be described as the birth certificate of modern ecology.

Eugene Odum identified several interrelated aspects of the environmental and "human predicament" and thus emerged as a major authority in the field. His brother Howard meanwhile, pursued an integrative and macroscopic approach based on systems theory, famously developing a whole energetics of ecosystems. In *Environment, Power and Society* (1971), he subsequently exposed and modelised the flows, phases, cycles and hierarchies of energy throughout the biosphere, drawing attention to the thermodynamics at work in nature and in its human management and concentration in agriculture, commerce, industry, urbanism and information.

For three decades, while refining his energy systems language (*energese*) and clearing the ground for several research fields such as ecological economics and engineering (his work notably inspired the Biosphere II experiment), Odum relentlessly demonstrated that the industrial era would sooner or later face the predicament of energy descent caused by the increased rarity of dense fossil energy, and that securing what he called a "prosperous way down" required a drastic, voluntary rethinking and reorganisation of societies, economics and territories.

References
Eugene & Howard Odum, *Fundamentals of Ecology*, Saunders, 1953.
Howard T. Odum, *Environment, Power and Society*, Wiley & Sons, 1971.
Howard T. Odum, & Elizabeth C. Odum, *A Prosperous Way Down: Principles and Policies,* University Press of Colorado, 2001.

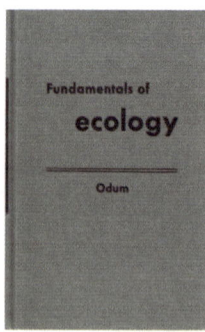

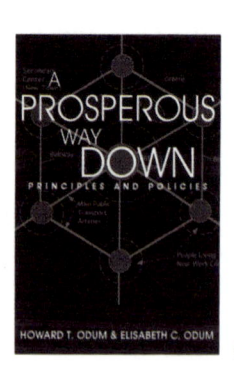

3
Fundamentals of Ecology,
Eugene Pleasants Odum, 1953.

4
Environment, Power and Society,
Howard Odum, 1970.

5
A Prosperous Way Down,
Howard and Elisabeth Odum, 2001.

– Qu'est-ce qu'on est bien ici !

SCHIZOPHRENIA
Martin Étienne, 2018.

E3

1

SIGNET • 451-W5767 • $1.50
THE LIMITS TO GROWTH
The headline-making report on the imminent global disaster facing humanity—and what we can do about it before time runs out. "One of the most important documents of our age!" —Anthony Lewis, The New York Times
DONELLA H. MEADOWS/DENNIS L. MEADOWS
JØRGEN RANDERS/WILLIAM W. BEHRENS III
A POTOMAC ASSOCIATES BOOK

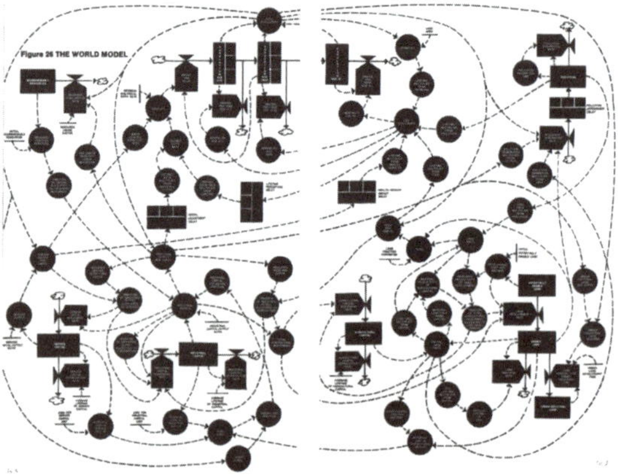
Figure 26 THE WORLD MODEL

2

3

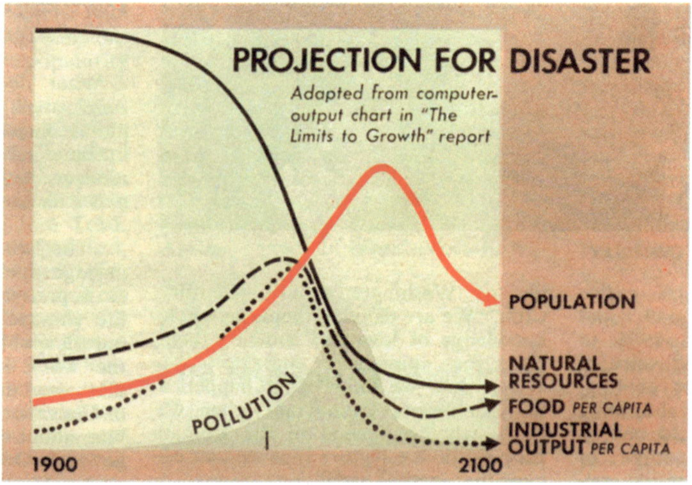
PROJECTION FOR DISASTER
Adapted from computer-output chart in "The Limits to Growth" report
POPULATION
NATURAL RESOURCES
FOOD PER CAPITA
INDUSTRIAL OUTPUT PER CAPITA
POLLUTION
1900
2100

4

E3

"Beyond the Limits is a book of stunning intelligence." —BARRY LOPEZ
BEYOND THE LIMITS
CONFRONTING GLOBAL COLLAPSE
ENVISIONING A SUSTAINABLE FUTURE
Sequel to the International bestseller THE LIMITS TO GROWTH
DONELLA H MEADOWS, DENNIS L MEADOWS, JØRGEN RANDERS

5

1972: "THE LIMITS TO GROWTH"

SYSTEMS DYNAMICS VERSUS ECONOMIST DOXA

In 1968, on the initiative of Aurelio Peccei (then one of the top managers at Fiat), a group of industrial managers, scientists and senior officials of international institutions, who all shared concerns about the state of the world and its problematic future, had its first meeting in Rome, and was thence baptised The Club of Rome. Two years later, its first major initiative was to commission a report from Jay Forrester's Lab of Systems Dynamics at MIT, where a team of 16 young post-docs, led by Dennis Meadows, took up the challenge. In two intense years of research, the team collected data from around the world on a number of essential parameters (demographics, consumption of natural resources, food per capita, industrial output, pollution and degradation of the environment) and fed them into a model (World 3) which mapped their evolution and interaction from the beginning of the 20th century and could make predictions on their global tendencies throughout the 21st century and thus allow experts to scrutinise the probable outcome of different scenarios.

In 1972, when they were published in a book called *The Limits to Growth*, the public was shocked. According to the model, the "standard run" scenario, where basic parameters would keep growing at a steady pace, the world was heading towards a global collapse — resulting from overshooting the "carrying capacity" of our planet — that would occur sometime around the first third or half of the 21st century. Moreover, with its other scenarios, the book showed that no single policy focused on checking just one parameter (demography, industrial output, etc.) could stop society from hitting the wall. Even the most optimistic scenario, which hypothetically doubled our planet's estimated reserves of natural resources, only delayed the collapse by one or two decades (and made it worse!). In other words, the main lessons drawn from the model were firstly, that there were indeed "limits to growth" which would soon assert themselves globally. The second lesson was that preventing this global collapse required a conscious and active transition based on a holistic approach (and not a mere technological fix of a particular parameter); and finally, that delay in engaging this shift would necessarily reduce and compromise, if not prevent altogether, the ability of mankind to cope with this predicament.

While the book was quickly discussed worldwide, and started to inspire other research, it immediately came under heavy fire from neoclassical economists who were eager to dismiss it as scientistic, interventionist and basically irrelevant. And indeed, as the two oil crises of the 1970s were over and the neoliberal ethos reasserted itself internationally, it all appeared as if the predictions of the first report to the Club of Rome had just been the effect of a very bad trip. But during their dry spell, its authors, refining their model and their data, did produce, in 1992 and 2004 respectively, two updates showing that, in line with their initial predictions, global economy had now grown well "beyond the limits".

References
Donella and Dennis Meadows, Jörgen Randers, William Behrens, *The Limits to Growth: A Report for the Club of Rome's Project for the Predicament of Mankind,* Chelsea Green, 1972.
D. and D. Meadows, J. Randers, *Beyond the Limits,* Chelsea Green, 1992.
D. and D. Meadows, J. Randers, *Limits to Growth: the 30 Year Update,* Chelsea Green, 2004.

1
***The Limits to Growth,* Donella H. & Dennis L. Meadows, Jørgen Randers, and William W. Behrens III, 1972.**

2
Systems dynamics: The World Model 3, *The Limits to Growth,* 1972.

3
A red line for an impending worldwide crisis: Donella H. & Dennis L. Meadows, Jørgen Randers, and William W. Behrens III.

from *Time Magazine* 24 January 1972

4
The team behind the *Limits to Growth,* with Jay Forrester, 1972.

5
Twenty-year update: *Beyond the limits,* Donella H. & Dennis L. Meadows and Jørgen Randers, 1992.

INHABITING A BIOREGION
Drawing by Jay Kinney for Murray Bookchin, "On Social Ecology",
CoEvolution Quarterly, 1981.

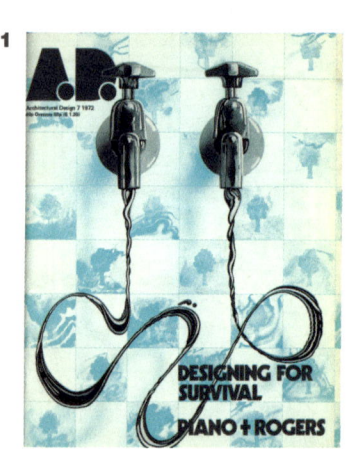

1

2

3

"BEYOND INDUSTRIAL TECHNOLOGY"

TOWARDS AN ALTERFUNCTIONALIST AGENDA

In July 1972, *Architectural Design* (AD) devoted an entire issue to the theme chosen by the RIBA for its annual conference: "Designing for Survival, Architects and the Environmental Crisis". The guest editor was Colin Moorcroft (b. 1947), a young architect who wrote the regular Eco-Tech column in AD. An omnivorous reader on biology, energy, technology, system-thinking and environmental issues since his teens, Moorcraft, while a student in architecture at the Architectural Association School, had also written a book, *Must the Seas Die?*, which was published that same year.

The title essay that Moorcraft contributed offered a very substantial and well-informed assessment of the deleterious effects of industrial technology, and especially of its massive application to the field of agriculture and natural resources, i.e. the so-called "Green Revolution". In retrospect, the clear-sightedness with which it described and articulated all these social and environmental effects seems strikingly prescient: energy and material cycles, biodiversity, climate change, metabolic rift, poverty and domination, rural exodus and migrations, planned obsolescence, etc.). Moorcraft's central charge was against "simplification". He wrote: "Until the rise of industrial technology the surface systems of this planet were continuously evolving into more complex and more stable form. We have reversed that trend and in a few decades have undone the work of millennia. The complexity and stability of the atmosphere, hydrosphere and biosphere have all been adversely affected."

In Moorcraft's view, high tech flights of fancy such as the cybernetic dream of a future freed from labour by automated lines of production, or the cabin ecology of closed systems and spaceships, *à la* Buckminster Fuller, were ultimately just dead ends. What was needed was to think "beyond industrial technology" and devise ad hoc "mediations between man and his environment" that would minimize entropy by emulating and stimulating the complexity of ecosystems. In his conclusion, Moorcraft delineated the three basic principles of this post-industrial technology:

1. Cooperation: "Each element should, wherever possible, be capable of performing more than one function and conversely each function should be performable in more than one way."
2. Integrity: those technologies should track, integrate and cycle the maximum of their inputs and "externalities", instead of simply aiming at some abstract "efficiency".
3. Flexibility: those techniques were to be sufficiently light, understandable and adaptable so that they could "respond to an extreme variety of social situations", evolve locally without requiring "an inflexible master plan", and "offer opportunities for self-servicing outside the technocratic system".

In all these respects, the systems had a lot to do with (and a lot to learn from) the vernacular knowledges and practices that the industrial and the Green Revolutions had eradicated.

Clearly, what this powerful essay pleaded for was a kind of alter-functionalism which, modelled on natural rather than industrial systems, stood in sharp contrast to the functionalist belief system adopted by the pioneers of the modern movement. To say that Moorcraft's views did not have much influence on the architectural milieu would be a gross understatement. Fortunately, they were not lost on another breed of designers.

References
Colin Moorcraft, "Designing for Survival", in *Architectural Design*, July 1972.

1, 2
"Designing for Survival", Colin Moorcraft, *Architectural Design*, July 1972.

3
Self Sufficiency Housing, *Homes & Cities*, Colin Moorcraft, 1982.

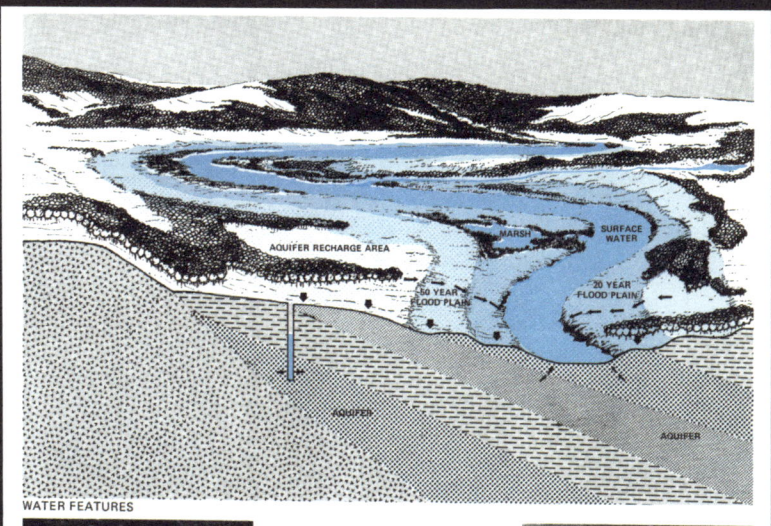

WATER FEATURES

SURFACE WATER

MARSHES

FLOODPLAINS

SITE GENERATED PLANNING
Design with Nature, Ian McHarg, 1969.

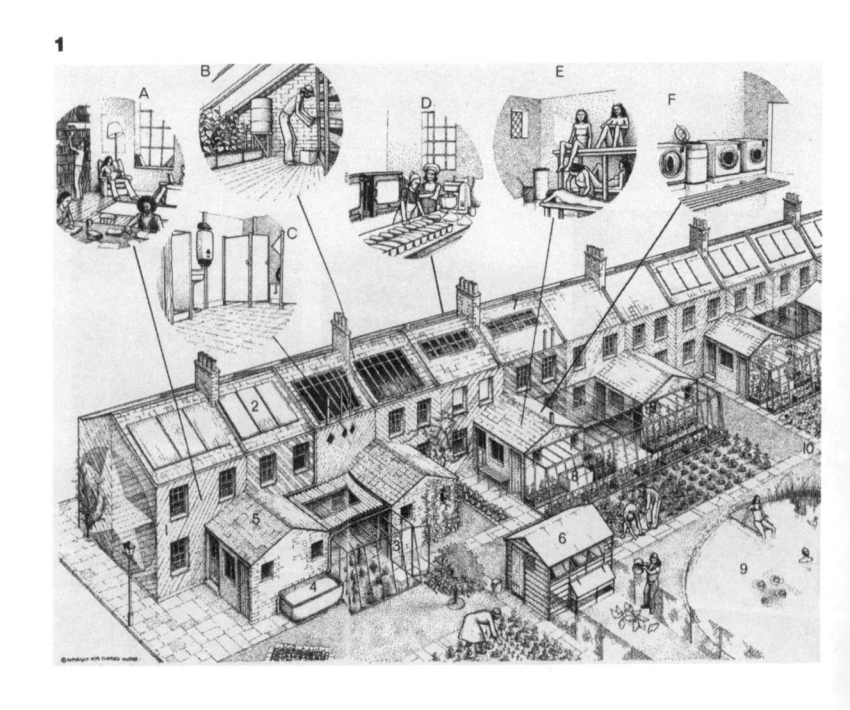

1
Autonomous Terrace,
Radical Technology, **1976.**
Clifford Harper

"SMALL IS BEAUTIFUL"

INTERMEDIATE, ALTERNATIVE
OR RADICAL TECHNOLOGIES

In the 1960s and 1970s, evidence of the environmental and energy crises led to a deep questioning of the goals, scales and dynamics of modern technology, as well as its social and political consequences. These were significant developments on Lewis Mumford or Jacques Ellul's early indictments of the "megamachine" or the "technological system". Most influential in this respect were Ernst Friedrich Schumacher's essays, collected in his famous *Small Is Beautiful: A Study of Economics as if People Mattered* (1973). An original and brilliant economist who was well aware of energy issues (having advised for two decades the British National Coal Board), Schumacher had also been, as a consultant in developing countries, a keen observer of the dire social disruptions forced onto those regions by technology transfers from the West. Against mass production systems and policies which privileged "man the producer" over "man the consumer", and had the effect of undermining local social structures, making people "footloose", he pleaded for "appropriate" or "intermediate" technologies: wiser methods and equipment, conducive to self-reliance and an "economy of permanence". Schumacher listed the following qualities that these tools need to have: "1. Cheap enough so that they are accessible to virtually everyone; 2. suitable for small-scale application; and 3. Compatible with man's need for creativity."

If this critique of technology, and re-evaluation of vernacular cultures, had already strong echoes and equivalents in the sphere of design — i.e. Victor Papanek and his *Design for the Real World: Human Ecology and Social Change* (1971) —, the early 1970s saw a blossoming of collective endeavors, more politically engaged, which were eager to push further the quest for self-sufficiency and autonomy. Most exemplary here is the nebula of authors who contributed to *Undercurrents*, the "magazine of alternative science and technology" launched in 1972 by Welsh activists Godfrey Boyle and Peter Harper, and eventually to the impressive compilation edited by the same. The book *Radical Technology* (1976) covered a huge range of fields from "Food-Shelter" to "Tools-Materials" and "Autonomy-Community". The book posed the questions: "How far can economic and resource autonomy practically be taken on a small scale? ... Does autonomy benefit only those who undertake it, or could it benefit society as a whole? What if everybody did it?"

References
Victor Papanek, *Design for the Real World: Human Ecology and Social Change*, Bantam Books, 1971.
Ernst Schumacher, *Small is Beautiful: A Study of Economics as if People Mattered*, Harper & Row, 1973.
Peter Harper and Godfrey Boyle, *Radical Technology*, Pantheon Books, 1976.

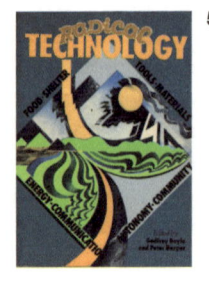

2, 3, 4, 5
Scaling down tools and technologies: *Small is Beautiful,* **1973,** *Farming for self-sufficiency,* **1973,** *Undercurrents no. 17 — Inner Technology Special Issue,* **August–September 1976,** *Radical Technology,* **1976.**

PROGRAMMATIC TAPESTRY, METROPOLITAN CROP ROTATION
Proposal for Parc de la Villette, OMA, Paris, 1982.

1

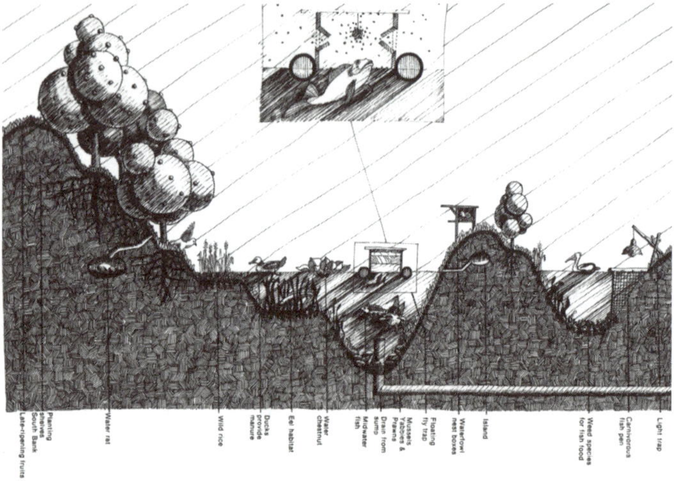
Late-ripening fruits
South Bank
Planting shelves
Water rat
Wild rice
Ducks provide manure
Eel habitat
Water chestnut
Midwater fish
Drain from sump
Mussels Yabbies & Prawns
Floating fly trap
Waterfowl nest boxes
Island
Weed species for fish food
Carnivorous fish pen
Light trap

2

PERMA-
CULTURE
ONE.
A Perennial Agriculture for Human Settlements
Bill Mollison & David Holmgren

3

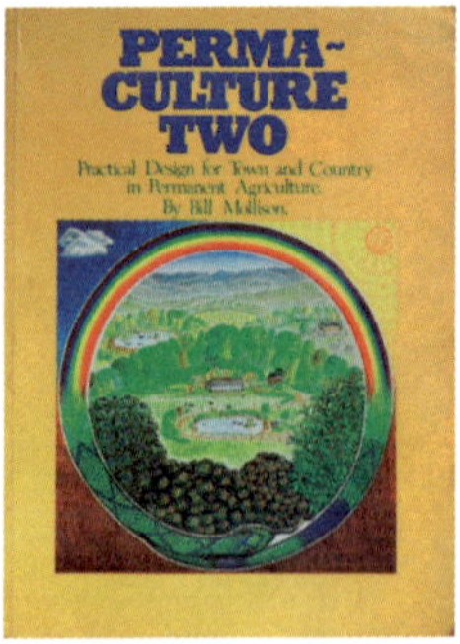
PERMA-
CULTURE
TWO
Practical Design for Town and Country
in Permanent Agriculture.
By Bill Mollison.

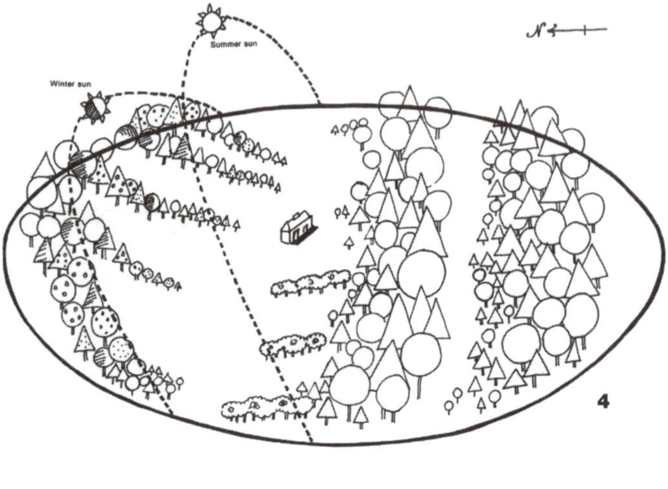
Summer sun
Winter sun
N

4

1978: "PERMACULTURE ONE"

A PERENNIAL AGRICULTURE FOR HUMAN SETTLEMENTS

After high school, David Holmgren, born in 1955 to a family of Australian middle-class radicals, spent a year hitch-hiking and working in construction jobs across the country. At the end of this period, in contrast to his initial plan which was to study architecture, the young man, eager to experiment with radical ways of articulating agriculture, ecology and landscape design, opted for the new programme of Environmental Design in Hobart, Tasmania. There, he soon connected with 30 years older Bill Mollison, a highly original figure who had evolved from fisherman, rabbit-hunter, bushman and lumberjack to a wildlife ecologist before studying bio-geography at Hobart where he subsequently taught environmental psychology. Teeming with ideas and extremely critical of industrial methods in agriculture, Mollison was also convinced of the uselessness of all forms of critique that would not be based on the definition and demonstration of viable alternatives.

Intense exchanges took place between the two men, to which the young Holmgren contributed a wide scientific curiosity and a propensity for sound and structured reflection. Alongside their experiments in organic horticulture and forestry, the study of indigenous agrosystems, and all the literature they mined on agronomy, anthropology and ecology, Holmgren specifically weaved two threads of concerns that were to take a growing importance in his work: firstly, system's energetics (especially Odum's) and secondly, the theory of design and site planning as evinced by Ian McHarg, Kevin Lynch, Christopher Alexander and others. The outcome of their collaboration was the idea of *permaculture* (a contraction of "permanent agriculture"), and its exposition in a dense essay, written by Holmgren, which the

21-year-old submitted as reference for his diploma design in 1976. Two years later, while Holmgren had set out to practically engage with the idea as both builder and gardener, Mollison published a revised version of the manuscript which immediately gained public attention and launched his career as spokesman for the permaculture movement.

In a nutshell, permaculture was explicitly conceived as a design approach to subsistence gardening, enabling households or communities to address the environmental predicament locally by stewarding self-sustaining sites providing for their basic needs. A most consistent implementation of the alter-functionalist principles delineated by Colin Moorcraft, its aim was to empower people to become the conscious and responsible *designers* (architects and gardeners) of the multispecies ecosystems of which they were a part, and thus turn their mode of living into an art of resilience and relative self-sufficiency modelled on perennial forest ecosystems. What was at stake was to evolve what Mollison and Holmgren called "consciously designed landscapes which mimic the patterns and relationships found in nature, while yielding an abundance of food, fibres and energy for provision of local needs." Indeed, by applying design thinking to the field of subsistence gardening and local self-sufficiency, permaculture was radically questioning the rationality of modern planning, and reformulating the principles of architecture.

References
Bill Mollison & David Holmgren, *Permaculture One: A Perennial Agriculture for Human Settlements*, Transworld Publishers, 1978.
Bill Mollison, *Permaculture Two: Practical Design for Town and Country in Permanent Agriculture,* Tagari Publications, 1979.

1
"Some elements of a Pond Polyculture ", Bill Mollison, David Holmgren, 1978.

2, 3
Permaculture One, Bill Mollison, David Holmgren, 1978 and *Permaculture Two*, Bill Mollison, 1979.

4
Climate planning, winter and summer sun angles decide for tree height in planting, *ibid.*

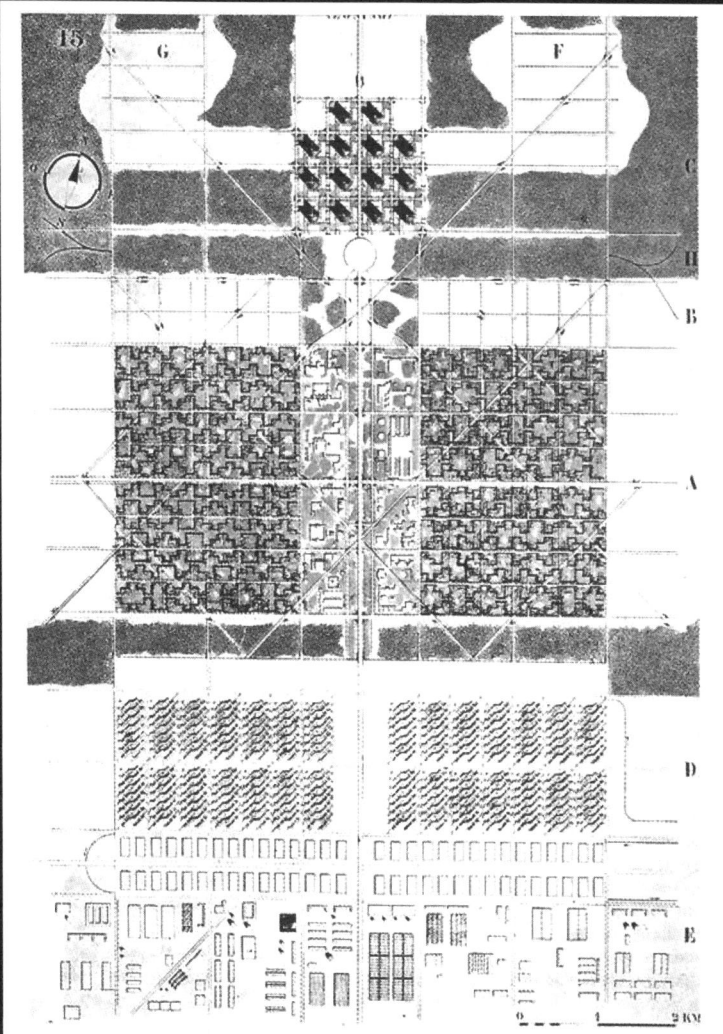

La planimetria della Ville Radieuse (Le Corbusier).

A, abitazioni; B, alberghi e ambasciate; C, città degli affari; D, industrie; E, industrie pesanti (fra le due i depositi generali e i docks); F, G, nuclei satelliti con caratteri speciali (per es., città degli studi, centro del governo, ecc.); H, stazione ferroviaria e aeroporto.

FUNCTIONALISM 1.0
Segregation of Functions, Radiant City Plan, Le Corbusier, 1922.

1

2

2008: "FUTURE SCENARIOS"

A USEFUL COMPASS FOR NAVIGATING THE 21ST CENTURY

Permaculture has certainly been, to date, one of the most consistent responses to the environmental alarm that has been raised so forcefully since the 1960s. However its political relevance beyond the domestic or micro-communal levels is often questioned. Indeed, it is a philosophy of doers who feel deeper urge to operate changes locally rather than "occupy Wall Street" or join the chorus of *indignados*. But this does not proceed from some kind of internal exile nor from an indifference to the larger picture. David Holmgren's numerous contributions to the understanding of the wider context are quite exemplary in this respect.

In 2008, shortly before the financial crisis, Holmgren published *Future Scenarios* which contrasted four common views of the future:

1. Techno-explosion — new sources of dense energy, space conquest, etc.
2. Techno-stability — sustainable development, solar panels, etc.
3. Collapse — a *Mad-Max* degringolade into survivalism.
4. Descent — a succession of crises and plateaus progressively unwinding the industrial era and leading to biological resources becoming ever more crucial.

As a keen reader of Howard Odum, Holmgren thinks that this last view, the most neglected by mainstream futurologists, is also the most plausible, and likely to be driven by the interaction of two simultaneous processes: peak oil (the rarefaction of dense energies with high transformity) and climate change (the progressive degradation of life conditions).

Since each of those two processes may unfold either slowly or rapidly in the coming decades, this opens up four possible combinations (4 quadrants) resulting in four different mid-term scenarios which Holmgren characterizes as:

1. Brown-Tech — slow energy descent and severe climate change, leading to strong and concentrated state interventionism.
2. Green-Tech — slow energy descent and mild climate change which allows a planned transition toward "sustainable development".
3. Earth-Steward — severe energy descent and mild climate change which is conducive to urban exodus and a new cult of local resilience.
4. Lifeboats — fast energy descent *and* severe climate change which is causing a devolution of large social organizations into a constellation of competing tribes, gangs and feudal lords.

In Holmgren's view, these scenarios are not mutually exclusive: they may well develop next to one another as all regions are not in the same situation, or even within one another like Russian dolls. In addition, they are also dynamically linked to one another: while Green-Tech would sooner or later morph into Earth-Steward, it is likely that Brown-Tech would finally explode into Lifeboats. Hence Holmgren's point: if permaculture is a practical philosophy of design that precisely corresponds to the Earth-Steward scenario — and would be marginal in Brown Tech — it is also highly relevant to Green Tech, and probably the best and most hopeful way of anticipating and preparing for a dark age of Lifeboats.

References
David Holmgren, *Future Scenarios: How Can Communities Adapt to Peak Oil and Climate Change,* Chelsea Green, 2009.

1
**ABCD Scenarios on the impact of energy transitions and converging crisis, Andrew Merritt, 2009.
EcoLabs / J.Boehnert**

2
Quadrants, *Future Scenarios,* David Holmgren, 2009.

A HORTICULTURE OF THE WELL TEMPERED ENVIRONMENT
Peach walls in Montreuil, Paris suburb, postal card, 19th century.

F.

REFRAMING THE PRACTICE & THEORY OF DESIGN

In which the mindful reader, reviewing the main concerns of permaculture — sparing efforts and energy *(utilitas)*, increasing resilience *(firmitas)* and managing worlds *(venustas)* — is led to wonder whether its proponents might not have evolved the most consistent theory of design since Vitruvius and Alberti, and the fiercest challenge to the alleged rationality at work in the spheres of agriculture, architecture and urbanism today.

SÉBASTIEN MAROT

166

MAJOR PRECEDENTS

KEYLINE PLAN AND NATURAL AGRICULTURE

"A stable, productive, and inherently beautiful landscape is perhaps the greatest material asset a society can inherit. Skill in landscape planning seems evident in some pre-literate agrarian cultures, but since the rise of professional skill rather than cultural traditions, landscape planning has focused on the urban environment and become cosmetic rather than utilitarian. Design of the productive rural landscape in modern industrial countries follows no stable traditional patterns nor any new rational science or art of landscape planning."
Permaculture One, op. cit.

Such was — and still is — the problem that Mollison and Holmgren intended to address with permaculture, which one could thus define as a practical and rational art (or design approach) to sustainably manage productive rural landscapes. Among the many references they explored and mobilised in their quest, two are particularly significant, both in themselves and in their combination.

The first one is the Keyline Plan developed in the 1950s by Australian cattle and sheep farmer Percival Alfred Yeomans (1905–1984): a rational approach to the design of farmlands, paying great attention to the retention and management of water, and the checking of soil erosion, the Keyline method envisioned agrosystems as basically defined by the fitting of eight "scales of permanence": climate, landform, water supply, roads, trees, permanent buildings, sub-divisional fences, and soil. With a view of attuning it to resilient horticulture, Mollison and Holmgren crucially improved this gradient by substituting "plant systems" to "trees" and by adding, immediately after, the scale of "microclimates". Indeed, permaculture can be seen as the art of managing productive landscapes as palettes of microclimates in the open: an agriculture of the well-tempered environment.
The second one is Masanobu Fukuoka's natural agriculture as exposed in his famous *The One Straw Revolution* (1975). A micro-

biologist by training, Fukuoka (1913–2008) was doing research on plant diseases when it struck him that science was essentially reductionist. He quit his laboratory and devoted his life to a small plot of land combining a mandarine grove and an acre of rice field. There, eager to minimise unnecessary work, he developed a system of wild or natural agriculture based on four principles: no ploughing as plants and microorganisms can do it themselves; no chemical fertiliser or prepared compost as the earth can take care of its own fertility; no tillage or herbicides but instead an enrolling of weeds in building soil fertility; and no dependence on chemicals. "Nature, left alone, is in perfect balance," he wrote. In a nutshell, Fukuoka evolved a clever calendar of plant associations and rotations (rice, clover and barley) where each acted as mulch or natural compost for the others. He also developed ingenious ways of perfecting simple traditional techniques such as broadcasting by rolling seeds in clay so as to avoid their being picked by birds.

A rebus for permaculture: designing and managing productive landscapes by emulating the ergonomics of nature.

References
Alfred P. Yeomans, *The Keyline Plan*, Waite & Bull, 1954.
Alfred P. Yeomans, *The City Forest*, Keyline Publishing, 1971.
Masanobu Fukuoka, *The One Straw Revolution*, 1975, (trans.), Rodale Press, 1978.

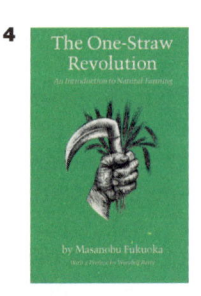

1
Farmland design: *The Australian Keyline Plan*, Percival Alfred Yeomans, 1954.

2
a, b, c, d: Yeomans' contour lines: primary valley, saddle valley form, ridge shapes and dams, 1958.

3
a, b, c, d: Sections of Fukuoka's argument, 1978.

4
Natural Farming: *The One-Straw Revolution*, Masanobu Fukuoka, 1978.

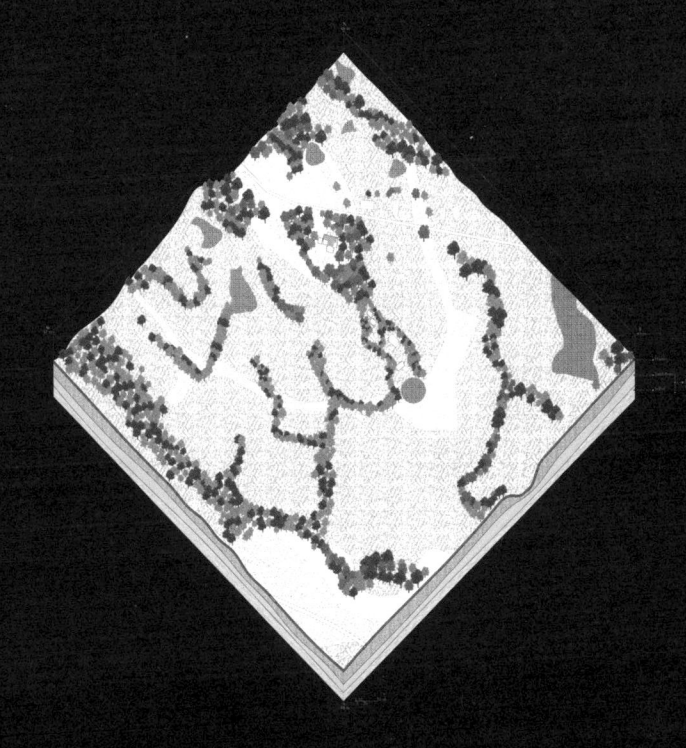

KEYLINE
Nevellan Farm, Percival Alfred Yeomans, New South Wales, Australia (Reda Erraziqi & Rose Hewins, 2019).

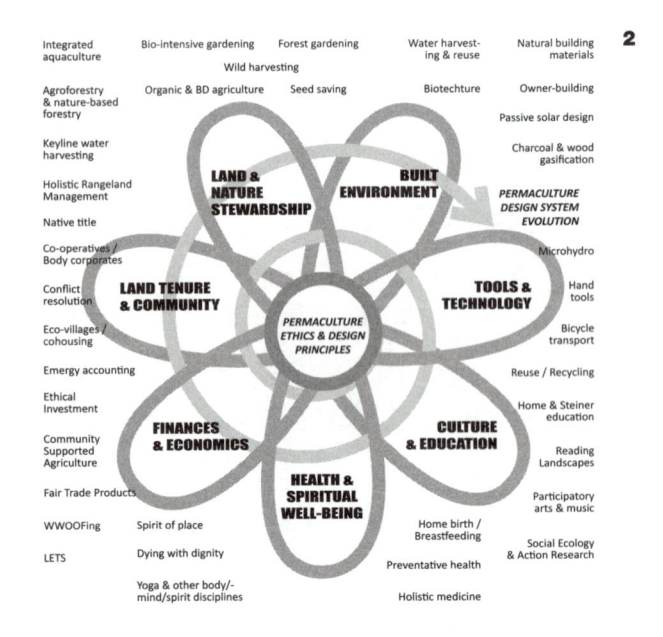

1
Permaculture ethics and design principles, David Holmgren, 2002.

2
The different fields of application: the Permaculture Flower, David Holmgren, 2002.

ETHIC AND DESIGN PRINCIPLES

A COMPREHENSIVE PHILOSOPHY
OF LOCAL RESILIENCE

David Holmgren did not just co-originate the concept of permaculture (with Bill Mollison) and write the first draft of the seminal book which launched it as "a perennial agriculture for human settlements". Twenty-five years later, by providing an overview of the basic ethical and design principles underlying the practice, he also equipped permaculture with a comprehensive philosophy and framed it, "beyond sustainability", as an art of living and thinking in the condition of energy descent which is now ours.

Holmgren sums up the structure and scope of this practical philosophy in two circular diagrams:

1. Three ethical principles — Care for the Earth, Care for the People and Fair Share (instead of, for instance, Liberté, Égalité, Fraternité) — form the core of a rose window of twelve design principles clearly distilling the basic precepts of a practical wisdom for a resilient agriculture and local land stewardship. The incomparable merit of this mnemotechnical figure is thus to gear its design theory onto the magnets of an ethics that explicitly incorporates the land and its non-human components: an ingenuous and powerful challenge to so-called architectural theory.
2. The permaculture flower illustrates, in a typically spiraling and holistic way, the overall relevance of those principles in all major fields of action: the biological field (land and nature stewardship) where the concept of permaculture originated; the built field (buildings, tools and technology) immediately adjacent; and the behavioural field comprising

culture & education, health, economics, and land tenure & community governance.

Perhaps designers who commonly brandish complexity as an excuse for aesthetic elusiveness will dismiss these diagrams as just another example of reductionist mandalas. They might want to reflect on the way Rem Koolhaas summed up the reaction of post-modern architects to his allegorical floating pool: "ignoring the spectacular decline of their profession, their own increasingly pathetic irrelevance… the limp suspense of their trite complexities, the dry state of their fabricated poetry, the agonies of their irrelevant sophistication, they complained that the pool was so bland, so rectilinear, so unadventurous, so boring… (In its ruthless simplicity, the pool threatened them – like a thermometer that might be inserted in their projects to take the temperature of their decadence)." In quite the same way, Holmgren's rose windows provide an excellent stethoscope to sound the common irrelevance, futility and cynicism of architecture and urbanism today.

References
Rem Koolhaas, *Delirious New York,* Oxford University Press, 1978.
David Holmgren, *Permaculture: Principles and Pathways Beyond Sustainability*, Holmgren Design Services, 2002.

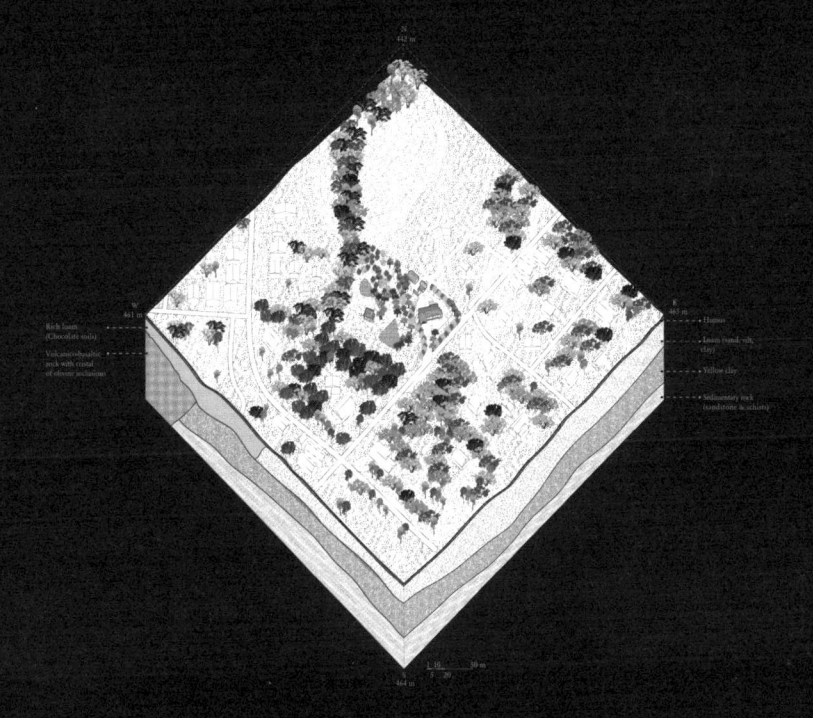

THE PIONNER'S GARDEN
Melliodora, Su Dennett & David Holmgren, Victoria, Australia (Reda Erraziqi & Rose Hewins, 2019).

1
Appearance of a developed permaculture view from sun sector, David Holmgren and Bill Mollison, 1978.

A RADICAL SITE-SPECIFICITY

"WORKING WITH WHAT
IS ALREADY THERE"

If Holmgren's rose window of ethical and design principles perfectly summarises the practical philosophy of permaculture, one might also discern within it four poetical intentions which define it not only as quasi-architecture but also as an art of managing worlds in times of energy descent and hence as an inspiration for both economy and architecture. In line with Holmgren's first principle ("Observe and Interact"), advising that no one should undertake any meaningful transformation of the site before at least a year (a whole cycle of seasons) of careful acquaintance with its physical, atmospheric and social conditions, the first of these intentions is a radically site-specific approach to design.

Patrick Whitefield, a British permaculturist who did much to transpose the principles of permaculture in temperate climates, puts much emphasis on this in the opening lines of his *Earth Care Manual* (2004): "The essence of permaculture is to work with what is already there: firstly to preserve what is best, secondly to enhance what is there, and lastly to introduce new things. This is a low-energy approach, making minimum changes for maximum effect, working in cooperation with both natural forces and human communities. Not only will solutions be different from region to region but from one locality to the next and even from one household to the next. Subtle differences of microclimate, soil and vegetation are taken into account, and so are the differences between the needs, preferences and lifestyles of different people."

This explains why permaculture should not be mistaken (as it has sometimes been) for a collection of technical or design recipes (raised beds, etc.) or frozen mandala-like patterns. By definition, its design principles are of general scope and must be transposed and balanced in each new context or situation according to the specifics of the place as well as the conditions and intentions of the people involved. As Dan Palmer stresses, a key to "making permaculture stronger" lies precisely in taking notice and advantage, of each site's unique qualities and latencies.

References
Patrick Whitefield, *Permaculture in a Nutshell*, Permanent Publications, 1993.
David Holmgren, *Melliodora, Hepburn Permaculture Gardens: Ten Years of Sustainable living,* Melliodora Publishing, 1995.
Patrick Whitefield, *The Earth Care Manual: A Permaculture Handbook for Britain and Other Temperate Climates*, Permanent Publications, 2004.
Patrick Whitefield, *How to Read the Landscape,* Permanent Publications, 2015.

2
Quixayá's village, Permaculture in the heart of Guatemala.
Lucas Wolf, 2015

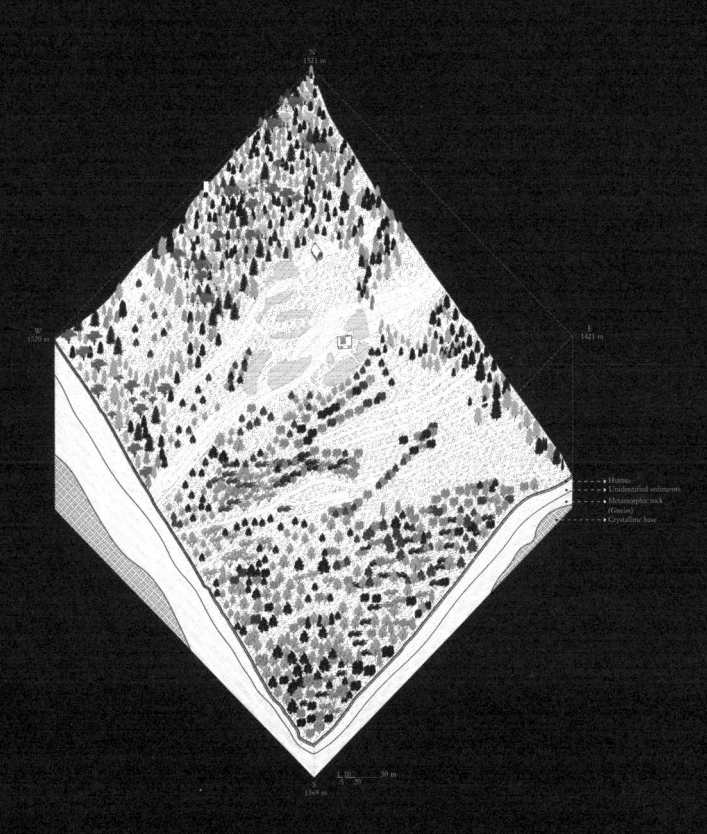

SLOPE
Krameterhof, Sepp Holzer, Austria (Reda Erraziqi & Rose Hewins, 2019).

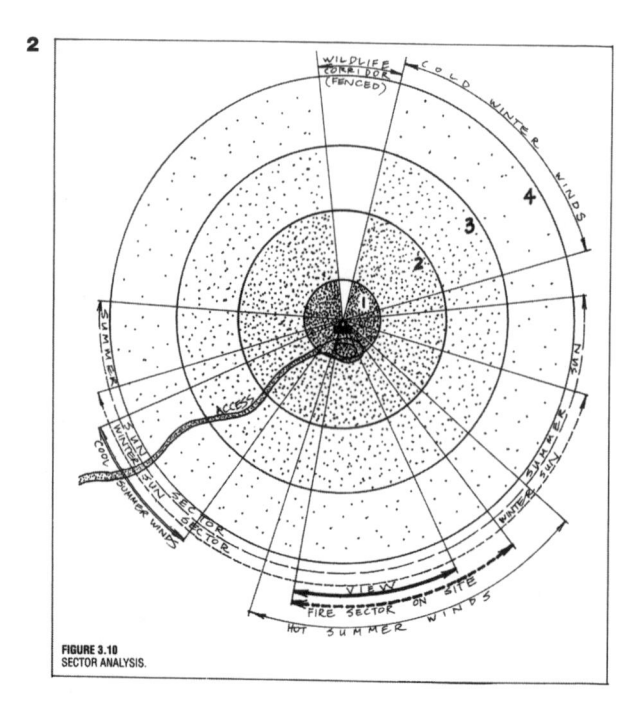

1
Intensity of Use/ Distance Relationship, David Holmgren and Bill Mollison 1978.

2
Zones and sectors regulate the location of particular plant species and structures, Bill Mollison, 1988.

A REFORMULATION OF SITE PLANNING

TOWARDS AN ERGONOMICS OF EARTH CARE AND RESILIENCE

"Traditional agriculture was labor intensive, industrial agriculture is energy intensive, and permaculture-designed system are information and design intensive." wrote David Holmgren. A basic intention of permaculture is thus to provide a reasoned and thoughtful, approach to site planning and stewardship to the effect of minimising both energy and labor in the long and short term. To ensure the processss is integrative and ergonomic, its key planning tools are zones, sectors and slopes (or elevation).

1. Zones in permaculture are basically an adaptation of Von Thünen's diagram (see panel B4) at the local and domestic scale: "whatever needs the most human attention and care should be placed nearest to the center of human activity" according to Patrick Whitefield. Starting from the house itself (Zone 0), one would typically get, in temperate climates, the following gradient: the home garden including intensive vegetable beds, salads and herbs, wall trained and other intensively grown fruit (Zone 1), orchards, poultry runs, housing for other animals, workshops and maincrop vegetables which require more space (Zone 2), field-scale crops and pasture, ideally integrated with productive water and small, intensively managed belts of woodland (Zone 3), rough grazing and woodland, where the value of yields for human use is relatively low (Zone 4), and finally wilderness or, say, "land where the interest of wild plants and animals take top priority" (Zone 5).

2. Sectors complicate the picture as they register the relationships of the site with influences coming from outside, such as wind, sunshine, flows of water, pollution, neighbors and views. "The principle of sectoring is to place things so that they have the best relationships with these influences". To

a great extent, as Whitefield emphasises, sectoring consists in working with microclimates (and soil structures) and taking the best advantage of their palette.
3. Elevation planning is "a matter of placing things in relation to the landform", refers Whitefield, and is crucial to make the best use of gravity in a world of low energy. Slopes, their gradients, shapes and curves, must be carefully taken into consideration when locating the different elements and uses of the farm so as to avoid soil erosion, spare efforts, and wisely manage water storage and flows.

References
Bill Mollison, *Permaculture, A Designer's Manual*, Tagari Publications, 1988.
Bill Mollison, *Introduction to Permaculture*, Tagari Publications, 1991.

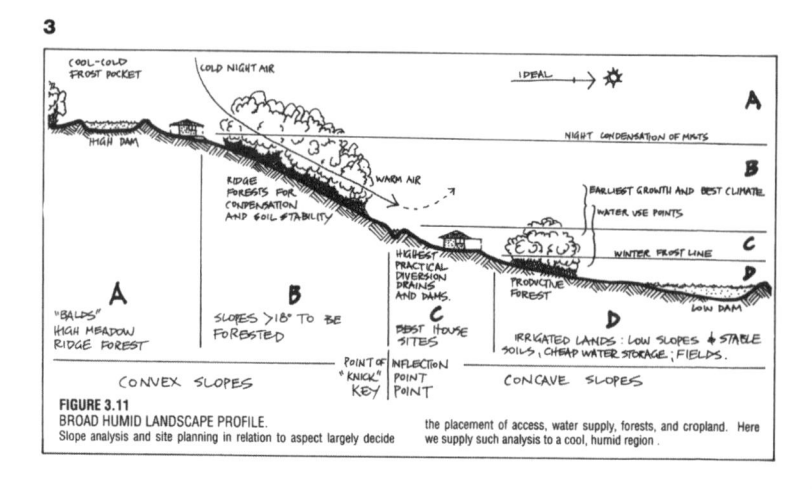

3
Ideal location of the elements in relation to the slope, Bill Mollison, 1988.

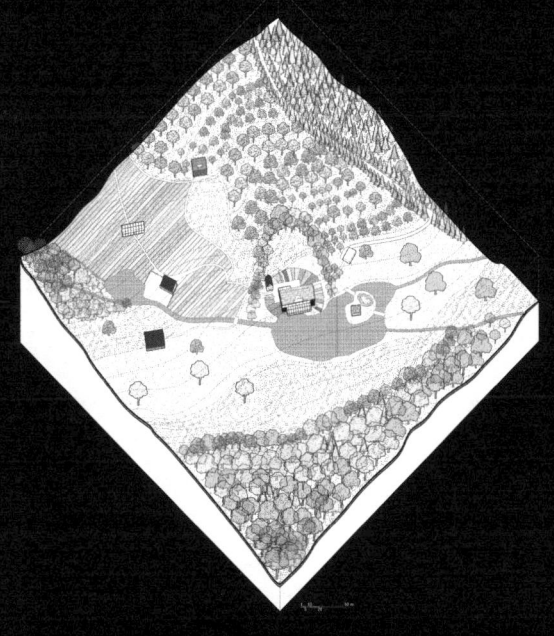

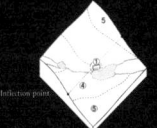

PERMACULTURE FARM MODEL
Reda Erraziqi & Rose Hewins, 2019.

1
Bec Hellouin's valley, with ponds, mounds and stakes, 2019.

2
A henhouse in a greenhouse — the use of volume at Bec Hellouin, 2019.

DEEPENING TERRITORIES

A MULTIDIMENSIONAL, VOLUMETRIC AND LAYERED POLYCULTURE

Permaculture's deepest intention is perfectly captured by Patrick Whitefield's notion of "multi-dimensional design": "Most agriculture, he writes, is virtually two-dimensional, consisting of low-growing field crops. Stacking introduces the third dimension, the vertical, succession works with the fourth dimension, time, while edge is about boundaries between different parts of the system."

Indeed, permaculture's model agrosystem is not the neat flatness of fields in annual herbaceous monoculture, but the volumetric perennial complexity of the forest. In other words, permaculture's site planning does not merely consist in carefully designing, organising, and distributing the plan or surface of the site. It is also, perhaps more importantly, a matter of managing and developing its whole section or depth, from the root system in the subsoil up to the aerial atmosphere of the canopy, so as to stimulate the most synergetic cohabitation and cooperation of multiple plant and animal species all along the cycle of seasons: a dense and resilient palette of interacting ecological niches. In that sense, permaculture is not just a quasi-architecture, or an extrapolation of architectural design into the extended field of agriculture: an articulated combination and layering of fruitful atmospheres and microclimates, it also entails a significant overcoming of the limitations of urbanism and urban design, which resulted from their growing divorce from (and subjection of) the spheres of non-human living organisms.

By designing and managing their sites as volumetric and storied polyculture, along scores of seasonal rotations and successions, permaculturists are clearly at the forefront of what should now be the major obsession of art: deepening territories so as to turn them into resilient worlds.

References
Joseph Russell Smith, *Tree Crops, A Permanent Agriculture*, Harcourt, Brace & Co, 1929.
Patrick Whitefield, *How to Make a Forest Garden*, Permanent Publications, 1996.
Perrine and Charles Hervé-Gruyer, *Permaculture: Guérir la Terre, Nourrir les hommes,* Actes Sud, 2014.
Perrine and Charles Hervé-Gruyer, *Vivre avec la terre,* vol. 2: "Cultures vivrières et forêts-jardins", Actes Sud, 2019.

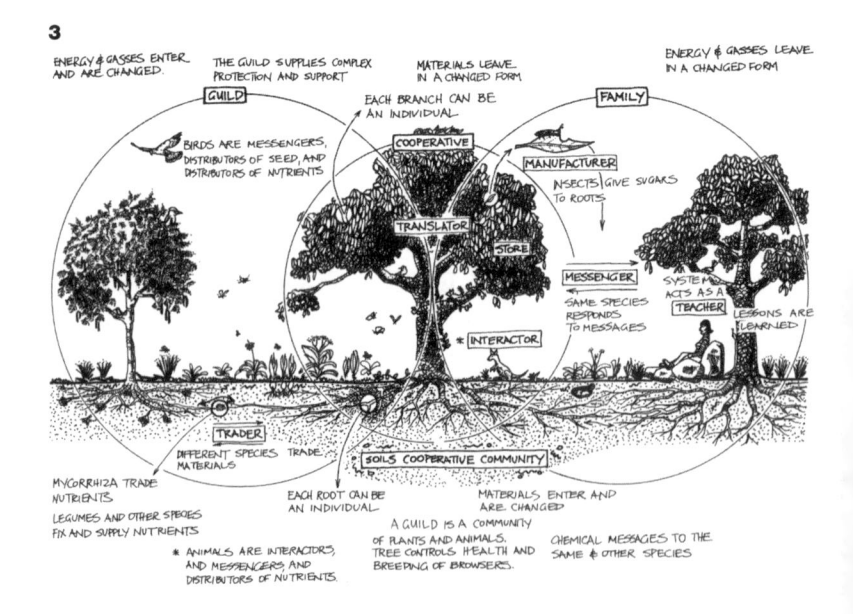

3
Trees in a whole System,
Bill Mollison, 1988.

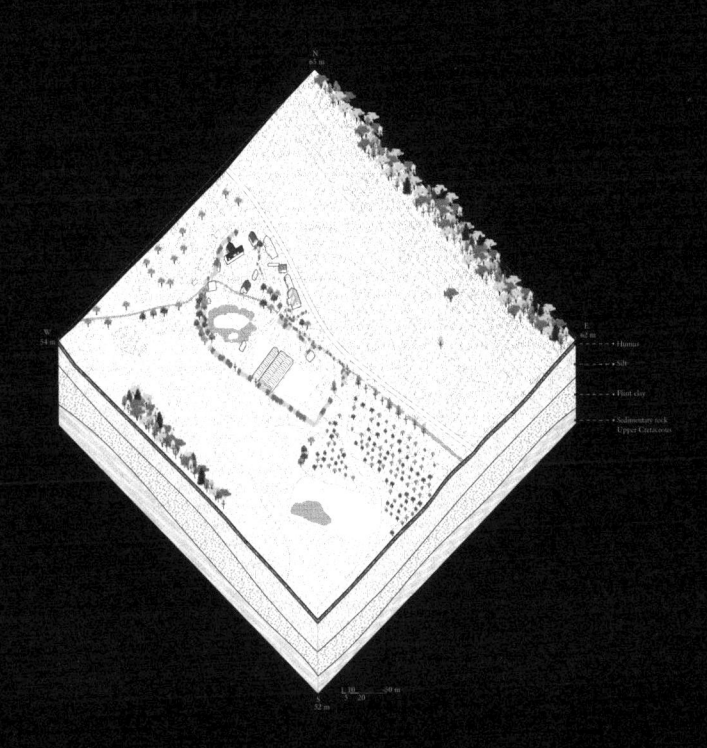

INTENSIVE MARKET GARDENING
Permaculture Farm Bec Hellouin, Perrine
and Charles Hervé-Gruyer, Normandy, France.
Reda Erraziqi & Rose Hewins, 2019.

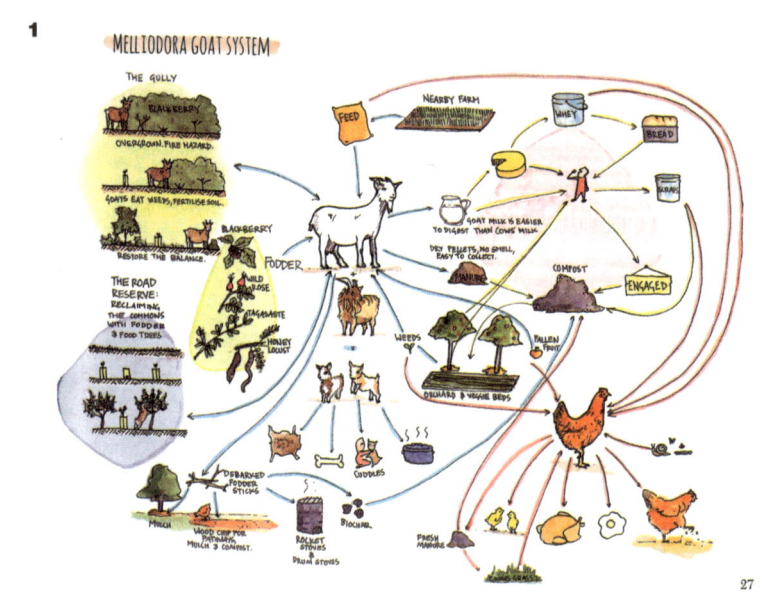

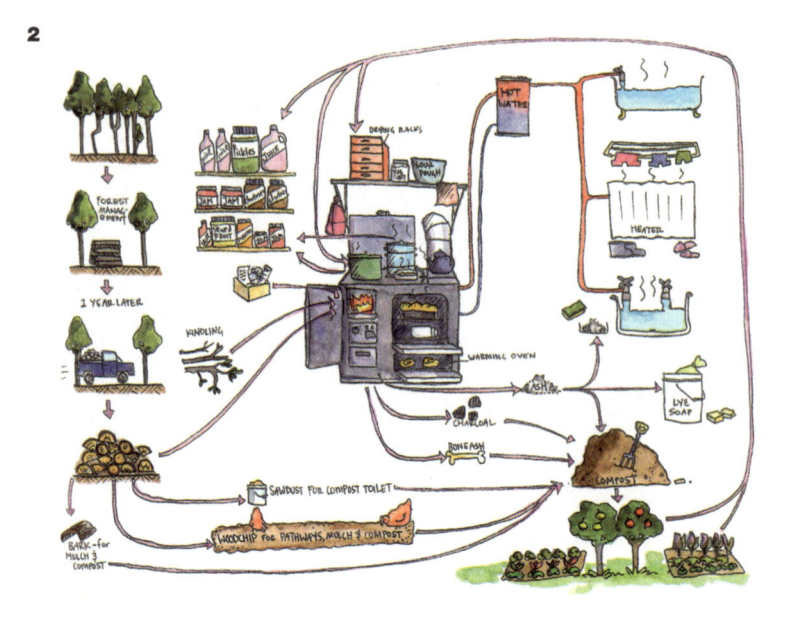

1
The Goat System, *RetroSuburbia*,
David Holmgren, 2018.

2
The House cycle, *ibid.*
Brenna Quinlan

AGRICULTURE AND ARCHITECTURE

RECONSIDERING URBANISM

TOWARDS AN ALTER-FUNCTIONALIST
DESIGN PRACTICE

Back in the 1920s, the great merit of functionalism was to claim that the rationality of architectural design cannot be immune to that of the larger historical context in which it operates. Taking stock of the fundamental changes brought about by a century of industrial revolution, it bluntly indexed the rationality of architecture and urbanism (their production, uses and aesthetics) on the scales, standards, processes and modes of production of modern industry. Since then much scorn and contempt has been poured upon functionalism by legions of Postmodernist architects — and particularly by the most purist and puritans among them, the so-called neo-rationalists — who all strived to sublimate their unquestioning adherence to the hyper-industrial ethos, thus chickening out from the real challenge raised by the Modernist pioneers.

Indeed, if architectural functionalism, in its Modernist version, must be radically questioned, it is not because of its celebration of function, but first and foremost because of its Modernism, i.e. because of the machinist, progressist, productivist and industrialist model it chose to embrace and emulate: a model which, based as it is on the ever growing availability of cheap and dense energy (fossil fuels), is obviously unsustainable on our finite planet and heading for collapse. Unfortunately, architects and urbanists, like most experts nowadays, still do not dare *believing* in what they *know*. Hence, they do not draw the consequences of the obvious by actively investigating other kinds of rationality and society.

By contrast, permaculture has been explicitly developed as a rational response to the environmental predicament and the condition of energy descent which are now ours. Instead of the principles at work in industrial processes (economies of scale, standardisation, segregation of functions, etc.), it thus strives to emulate and learn from those which operate in natural ecosystems (such as cooperation, complexity, integration, resilience, flexibility, etc.) and has evolved an alter-functionalist approach to design where indeed, according to Colin Moorcraft: "Each element should, wherever possible, be capable of performing more than one function and conversely each function should be performable in more than one way." (See panel E4).

References
Sébastien Marot, "L'Envers du décor", in Augustin Rosenstiehl (ed.), *Capital Agricole: chantiers pour une ville cultivée*, Editions du Pavillon de l'Arsenal, 2018.

3

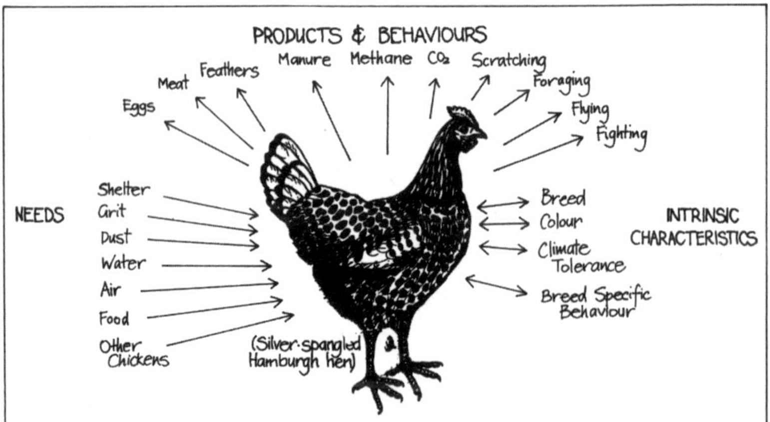

FIGURE 3.1
PRODUCTS AND BEHAVIOURS OF A HEN.
Analysis of these inputs and outputs are critical to self–governing design. A deficit in inputs creates *work*, whereas a deficit in output use creates *pollution*.

3
Permaculture Chicken,
Bill Mollison, 1988.

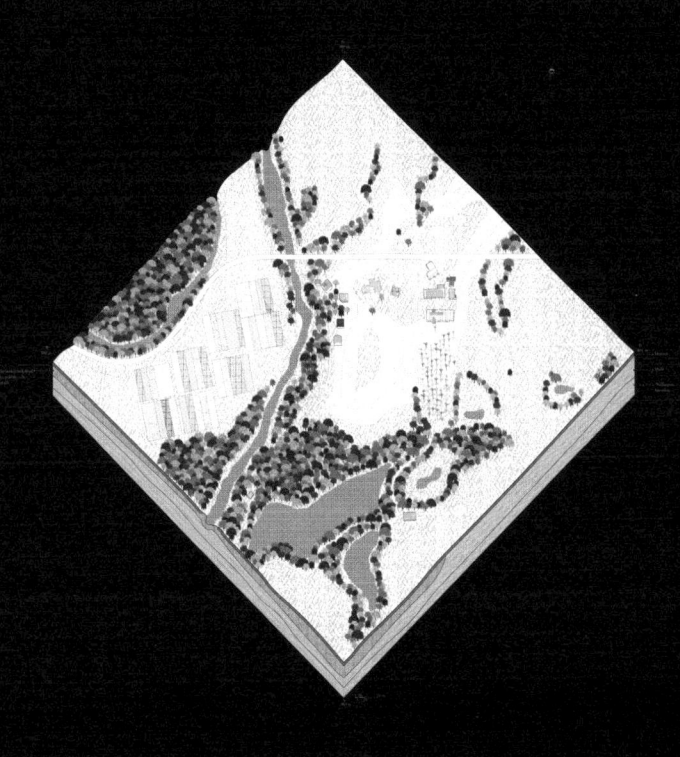

PERENNIAL POLYCULTURE
Land Institute, Wes Jackson, Salina, Kansas, USA.
Reda Erraziqi & Rose Hewins 2019.

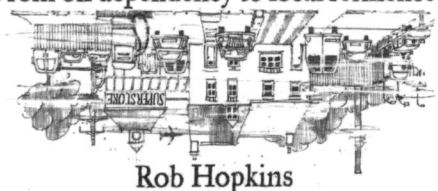

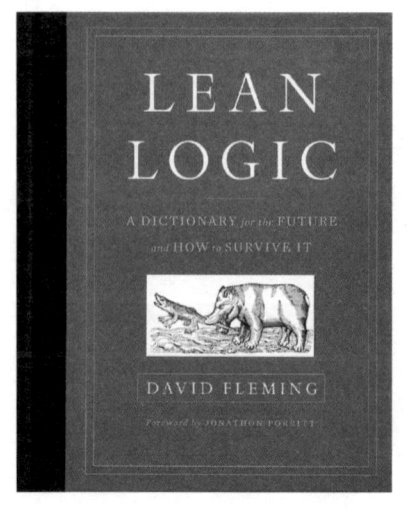

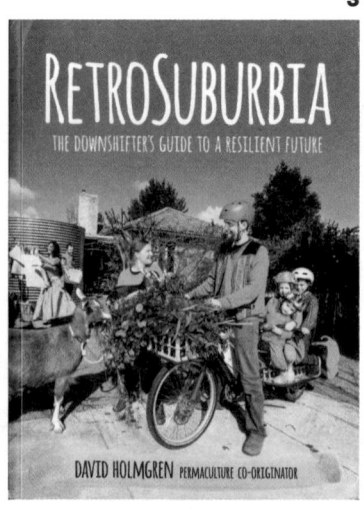

1
The transition handbook,
Rob Hopkins, 2008.

2
Lean Logic: A Dictionary for the Future and How to Survive It,
David Fleming, 2016.

3
RetroSuburbia, David Holmgren, 2018.

SUB-URBANISM?

WISE APPROACH, NOT SMART RECIPES

"Well, all that sounds very nice and gentle, but do you really want to make us believe that permaculture holds the key to feeding the billions that now swarm on our planet and keep packing in cities or suburbia everywhere? Get real! We don't buy it!". This is roughly what rationalist planners and problem-solvers will object with: they don't buy it… Well, perhaps that's because they can't: Permaculture design is not a political commodity and it is not on sale as a ready-made recipe that can be instantly expanded at any scale, regardless of the situation and the local particulars.

Indeed, permaculture is neither plan nor strategy. It is a practical philosophy that cannot be acquired but may be trialed by people in the habit of taking care of themselves, their surroundings and their communities, at least as much as imagining (and selling) global solutions for the world. In other words, permaculture is not *smart*, a word that has become the distinctive brand of everything which is patently *unwise*.

But this certainly does not mean that permaculture disregards the urban or suburban contexts in which most people live today, and that its approach would be irrelevant if one is to address the necessary evolution of urbanised territories. Quite to the contrary, the need and ways of increasing the local resilience and autarchy of urban communities, which were consubstantial to the very idea of permaculture right from the start, have also been the central subject of major developments on the part of key thinkers in the field.

One of them is the famous Transition Town movement that British permaculturist Rob Hopkins (b. 1968) initiated almost two decades ago. Combining David Holmgren's *Principles and Pathways* with David Fleming's lessons in "Lean Logic" for the patient rebuilding of local communities, this movement, first launched in two towns (Kinsale, Cork County, Ireland, and Totnes, Devon, England), advocates a bottom-up dynamic of spontaneous eco-municipalism which is now being emulated by local associations of citizens in a growing number of places — from villages to much larger cities — both in Europe and other continents.

But what about suburbia, those nondescript products of cheap energy and market economy that have been built over the past decades? Places where one can't rely on any legacy of public space patterns and communal living habits? This is precisely the topic addressed by David Holmgren in his last and most ambitious work, *RetroSuburbia: A Downshifter's Guide to a Resilient Future*. Illustrated by a bounty of examples drawn from local experiments led by permaculture pioneers in the Melbourne Area, the book suggests that typical suburbs, with their right balance of people and cultivable plots — why would you keep farming lawns instead of food? — might well be — much more than both distant rural areas and dense metropolises — the ideal breeding grounds for incipient and resilient communities… All this is a far cry from "smart cities" and "smart agriculture", which permaculture designers wisely abandon to their self-appointed champions and guinea pigs.

References
Rob Hopkins, *The Transition Handbook: From Oil Dependency to Local Resilience*, Chelsea Green, 2008.
David Fleming, *Lean Logic, A Dictionary for the Future and How to Survive It,* Chelsea Green, 2016.
David Holmgren, *RetroSuburbia: A Downshifter's Guide to a Resilient Future*, Melliodora Publishing, 2018.

4
Retrofit suburban home and garden, Beaconsfield, WA Australia, David Holmgren, 2018.

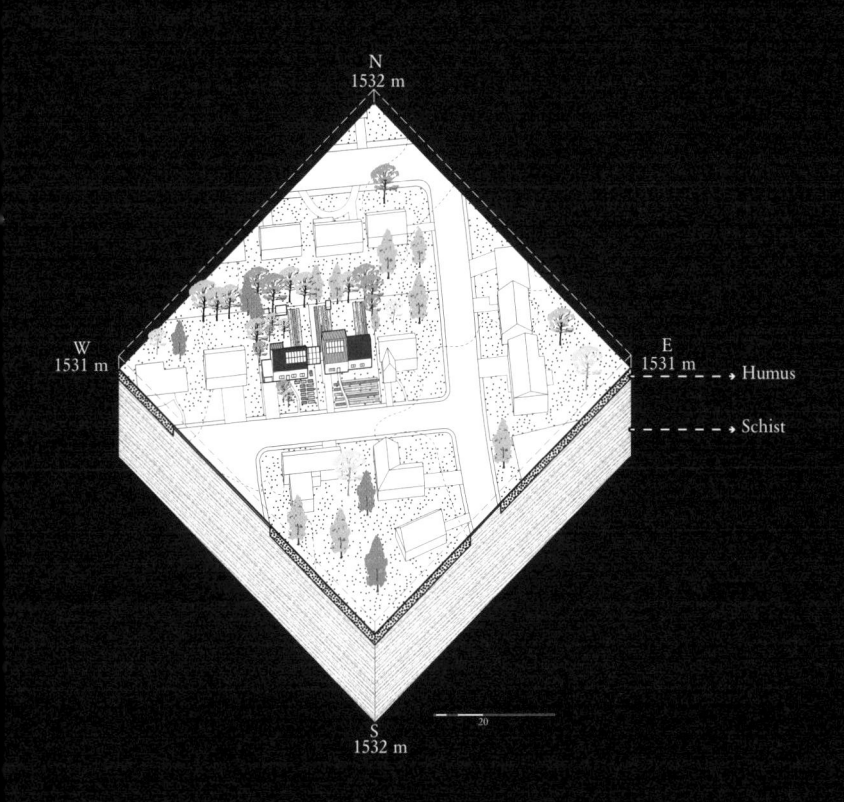

SUBURBIA
**Sunspot Urban Farm, Amy & Rod Adams, Fort Collins, Colorado USA.
Reda Erraziqi & Rose Hewins, 2019.**

G.

URBI ET ORBI

FOUR COMPETING NARRATIVES ON THE FUTURE RELATIONSHIP OF CITY AND COUNTRY

In which the now informed reader, equipped with a reasonably good rearview mirror on to the parallel histories of agriculture, architecture and urbanism, is finally introduced to a wind rose representing opposite scenarios in the type of relationship that city and countryside might develop in the near future, and gently invited to wonder which one (or two) of them she or he, in good conscience, could actively endorse.

SÉBASTIEN MAROT

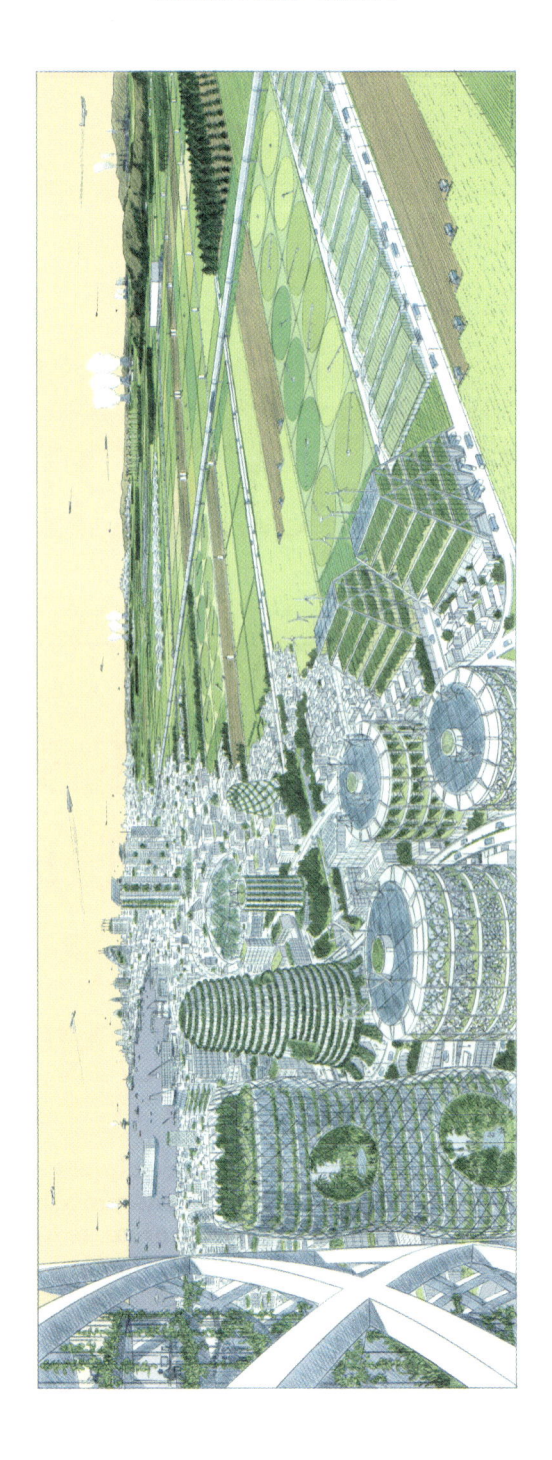

INCORPORATION

THE HIGHLY CAPITALISTIC METROPOLIS ABSORBS AGRICULTURE

What if the industrialisation of agriculture and its subjection to capitalism were logically leading to its urbanisation, or *incorporation* by the metropolis? Such is more or less the common accelerationist belief of those who, confronted with the dire environmental consequences of industrialised agriculture, imagine that the remedy is in the poison, and that only a flight forward into high-tech innovation and concentration may hold the key to a globally livable future. Mega glasshouses, vertical farms, high-rise feedlot buildings: thanks to the breakthrough and disruptive technologies of soilless culture, hydroponics and closed-system recycling, agricultural productions liberate their vast outlying peri-urban footprints to concentrate into biological reactors or fast-breeders, agri-buildings of agri-cities which turn synanthropic plant and animal species into cohabitants of the metropolitan Noah's Arch.

In this perspective, much embraced by the champions of eco-modernism, eco-pragmatism and agri-tecture (who are also clearly experts in hybrid linguistics), the metropolis is obviously envisioned not just as the *manifest destiny* of mankind, but also as the ultimate condition of our whole biosphere. Meanwhile, the dense city acts as a control tower surveying Countryside 2.0 consisting of a grid of robotised *latifundias*, interspersed with patches of productive forests, mines, natural preserves, and escapist leisure resorts, all scientifically managed by an army of experts. This ethos of concentration is well expressed by Stewart Brand in his *Whole Earth Discipline* (2008): "One emergent principle might be that deleterious elements should be concentrated. Concentrating people in cities is good. Concentrating energy waste products like nuclear spent fuel in casks is an improvement over distributing the greenhouse gases from spent coal and oil in the atmosphere. Concentrating our sources of food and fibre into high-yield agriculture, tree plantations, and mariculture frees up more wildland and wild ocean to carry out their expert Gaian tasks."

Many architects (to say nothing of engineers) seem tempted by this flight forward into what critic Peder Anker calls "cabin ecology", and dream (like its prophets, such as Buckminster Fuller) to precipitate the metabolism of calories in systems and circuits as closed, looped and controlled as possible. The term incorporation connotes the sur-rationalist absorption of agriculture by architectural and urban engineering as well as its ultimate subjection to the economic models of concentrated investment and management of hyper-capitalism.

SÉBASTIEN MAROT

NEGOTIATION

AGRICULTURE BECOMES
AN INTEGRAL COMPONENT
OF URBAN EXTENSIONS

This is the latent narrative of what we might call agricultural urbanism (in counterpoint to urban agriculture). Cities and metropolises take up spaces and species of agricultural production as integral components in the design of their margins and extensions. In this perspective, which challenges the modern demarcation line between urban, natural and agricultural zones, the latent capacities of agriculture, husbandry, horticulture and forestry to evolve arenas of consociation are hired by planning to foster an evolution of urban forms, syntaxes and modes of production. Park-orchards or park nurseries, market-gardens, housing developments, open campuses which mix education, agroecology and various activities, eco- and agro-districts, greenbelts or corridors of agroforestry, etc.: the list goes on and on of the new hybrid species that combine the best interests of cities and agriculture. These counter the deleterious dynamics of the metabolic rift between city and country and might also erode the persistent frontier between main job, secondary occupation and leisure activities.

Whereas this scenario may appear to be in its infancy today, it can claim some precedent in contemporary agro-ecology, and indeed there is a whole history of jurisprudence in the tradition and models of pre-modernist urban design: the agro-urban ideas and experiments that once converged around the concept of *civic design*. Weren't Olmsted's park systems, Howard's garden cities, Migge's Siedlungen, Geddes' Biopolis, Wright's Broadacre, etc., attempts at defining the figures and structures of an *agropolitan* future (to use a term that geographer John Friedman coined to describe certain regions of Asia)? Might it be time to resume their efforts by devising new contracts at all scales, "new deals", new forms of negotiation between urban and rural practices, that could restructure and give resilience to the *citta diffusa* that has spread and keeps spreading over entire regions?

Unsurprisingly, several of today's most influential approaches and trends in urban design, such as "landscape urbanism" or "ecological urbanism", more or less embrace this narrative of negotiation. They thus promote the idea of a horizontal metropolis which, far from containing and densifying the city against a backdrop of nature and agriculture, strive on the contrary to integrate and nurture the latter within the metropolitan fabric and field. Whether this narrative will succeed in de-simplifying urbanism and evolving more varied and polycultural syntaxes of coexistence, or will be merely hijacked as an alibi for the greedy and relentless growth of urbanisation, is an important, open question.

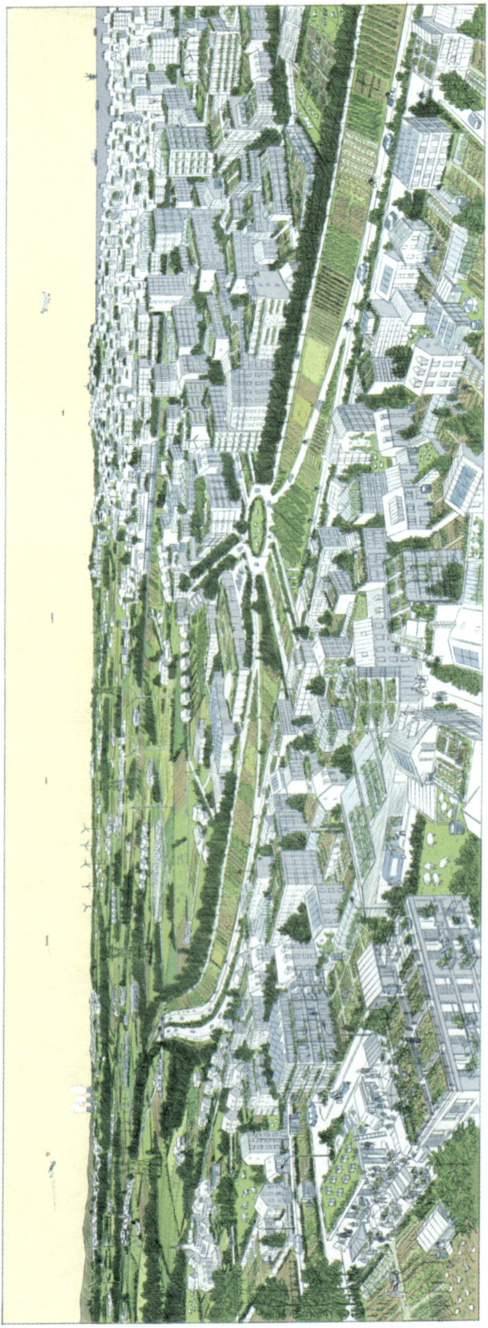

INFILTRATION

AGRICULTURE AND HORTICULTURE INVADE THE CITY

There is an underlying narrative to the work of those who take advantage of the neglected surfaces of cities and metropolises — such as roofs, terraces, vacant lots, median strips or sidewalks — to reintroduce horticulture and feeder gardening within the urban landscape; but also those who, reviving the practices of market gardening, build-up local networks that bypass the circuits of large-scale food business and retail. Without undermining the logic and realities of the urban condition, but rather by exploiting the latter's numerous niches, gaps and discrepancies, these varied initiatives take hold of food cultivation and consumption (and of their reintegration in local or short supply chain) as a means of building up collectives and solidarity-based practices in the uprooted territories of the metropolis.

Whether they proceed through direct integration with the fabric of the city, or through subscription to mixed-farming ventures and co-ops in the hinterland, these approaches all tend more or less to stimulate a higher degree of local interaction in urban territories which might evolve into a constellation of commons. Although it may be encouraged or faked by local authorities, infiltration is essentially a bottom-up phenomenon, an opportunistic and ad-hoc logic of self-organisation that does not pertain to planning or urbanism but blossoms here and there, like weeds, in the faults and gaps of urban territories. However, in contexts of severe economic decline or breakdown — such as the ones faced by La Havana (and Cuba in general) during the Special Period, or the City of Detroit after the collapse of its automobile industry, this phenomenon may obviously take on the dimensions of a landslide and significant recapture of urban plots by individual or collective food cultivation practices.

Since economic and energy crises are likely to strike a growing number of large metropolises and urban regions in the near future, and expand the amount of fallow urban areas, one may expect this scenario of infiltration (i.e. unplanned agricultural reclaiming of urban ecosystems and their suburban extensions) to become less and less like acupuncture and increasingly spread over larger metropolitan territories where it would evolve a variety of "rurban" fabrics, forms and syntaxes. How those will coexist with the palimpsests of surviving species and figures of the metropolis, and their struggle for existence, is anyone's guess.

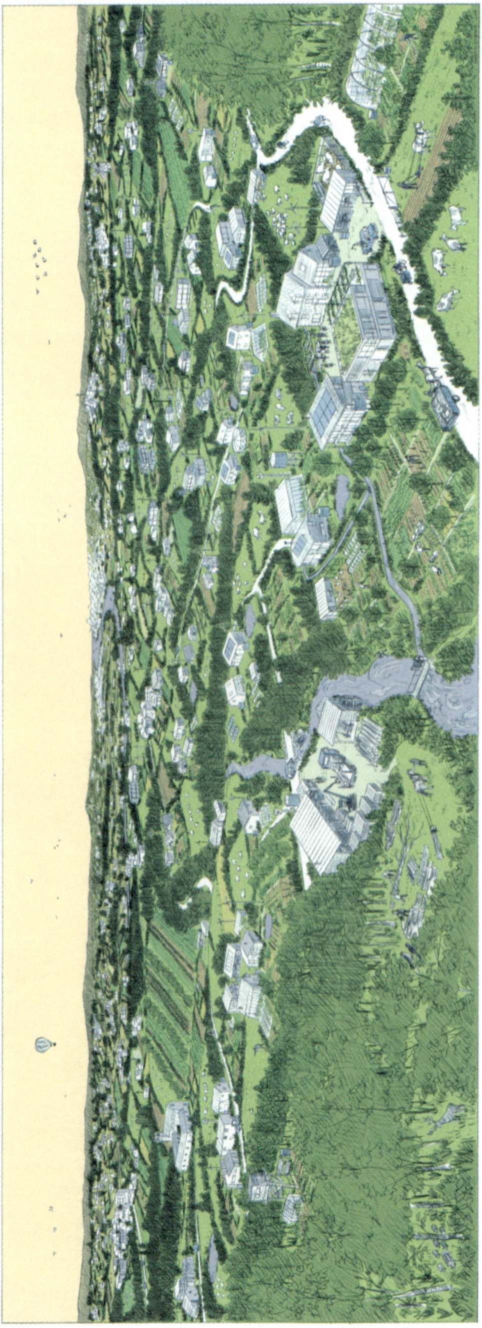

SECESSION

"IL FAUT CONSTRUIRE L'HACIENDA"

This is the more radical perspective of those who question the current hegemony of metropolitics, and hence the ability of urbanism to organise and maintain the eco-political conditions of resilient and satisfying worlds. From the overwhelming evidence gathered on the dire environmental, climatic, energetic and social consequences of consumer society and capitalistic concentration — of which metropolises and their touristic satellites are both the magnets and the most obvious products — the critics of the current politics of urban governance conclude that metropolitan territories are fundamentally unsustainable, doomed to collapse sooner or later, and that what is needed is for communities, by freeing themselves from their orbit and modes of "governance", to anticipate (if not accelerate) their progressive marginalisation and dismantling.

In this narrative of decentralisation, geared toward building means to achieve a significant degree of local autonomy, the principles of coexistence and the techniques of design and cultivation that enable people to tend a living landscape, a resilient community of interdependent humans, plants and animals, *supplant* urbanism. Alongside several other movements hinted at in this exhibition, from agrarianism to libertarian municipalism, permaculture is among the most disciplined expression of the agenda that would turn territories into confederations of self-managed communes or worlds.

Designating these experiments of non-urban foundation or re-foundation as *secession* may seem excessive. Many of these experiments are not necessarily framed as the antithesis to the metropolitan ethos but sometimes as simple offshoots or havens of "transition". Most, of course, must accept a certain compromise or *modus vivendi* with the rules and mechanisms of metropolitan governance. Besides, all of them may be more or less tolerated as "enclaves", or even hijacked and promoted as the prodigal offspring of a metropolis always eager to absorb contradiction by celebrating its own ecumenism. But three things must be here underlined. Firstly, there is a growing conviction with which these initiatives are dissociating themselves from the narrative of urbanisation as the manifest destiny of humankind. Secondly, there is a strong curiosity amongst participants in these initiatives in how to learn from one another, which turns them into the most active and prospective research centres. Finally, there is an intelligence and energy that participants manage to draw from the positive faith (or at least from the suspension of disbelief) that other natural covenants are eminently desirable, possibly achievable, and absolutely necessary.

In other words, what unites them in their very diversity, is their collective intuition that salvaging the idea of *civitas*, and giving it a new meaning, now badly requires a sub-version of and an exodus from the metropolis.

PUBLICATION

Editorial Concept
Sébastien Marot
Texts
Sébastien Marot, and Matthieu Calame (section C)
Illustrations
Martin Etienne (section G)
Research Assistance
Paul de Greslan (curatorial assistant), Gaétan Amossé, Raphaël Bach and Paul Bouet
Support
Observatoire de la Condition Suburbaine (OCS), Research Lab of the École Nationale Supérieure d'Architecture de la Ville et des Territoires in Paris Est
Copy Editing and Proofreading
Tim Abrahams
Editorial Coordination
projecto editorial
Publication Assistance
Carla Cardoso
Graphic Concept and Design
Marco Balesteros (Letra)
Design Assistance
Pedro Sousa, Robin Guillemin, Marion Cachon, Romain Guillo, Tiphaine Voix
Color Separation
Estudio Polígrafa / Carlos J. Santos
Printing and Binding
Gráficas 94, Barcelona

© of the images: the authors
© of the texts and translations: the authors
© of this edition: Lisbon Architecture Triennale, Lisbon; Polígrafa, Barcelona

ISBN [volume 2]: 978-84-343-1389-7
ISBN [set of 5 volumes]: 978-84-343-1393-4
Legal deposit: B. 39619 - 2019
New print run: 1.000 (2nd edition - February 2020)

Acknowledgements

For their inspiration, conversation, suggestions, challenges or critical comments: Pier-Vittorio Aureli, Matthieu Calame, Su Dennett, David Holmgren, Rob Hopkins, Wes Jackson, Rem Koolhaas, Dennis Meadows, Colin Moorcraft and Carolyn Steel.

And also: Armelle Antier, Joséphine Billey, Christophe Bonneuil, Jacques Delamarre, Sophie Deramond, Roland Freymond, Éric Lapierre, Romain Leonelli, Oscard Morand, Paule Pointereau, Marco Rampini, Augustin Rosenstiehl, Matthew Skjonsberg, Cyril Veillon, and students from the master Architecture & Expérience at the École d'Architecture de la Ville et des Territoires Paris Est: Nils Bailly, Reda Erraziqi, Rose Hewins and Pauline Kinsky.

All rights reserved. No part of this publication may be reproduced, stored in a retrieval system or transmitted, in any form or by any means, electronic, mechanical, photocopying, recording or otherwise, without permission in writing from the Publisher.

Every effort has been made to contact the owners and photographers of the images we're publishing here. Anyone having further information concerning copyright holders is asked to contact the publishing house (info@poligrafa.com) so this information can be included in future printings.

Image credits

© Abbaye Notre-Dame de Sénanque, Centre des Monuments Nationaux: p. 110 (3). © Albert Pope and Jesús Vassallo: pp. 40 (1, 2), 42 (3). © American Museum of Natural History, Natural History Press: p. 151. © Andrea Branzi: p. 72 (5, 6). Archiv der Max-Planck-Gesellschaft, Berlin: p. 90 (5). Archives du Calvados (H/3253/1-H/3253/4): p. 80 (2). Bibliothèque nationale de France, département Réserve des livres rares (S-1032): pp. 78 (4, 5), 76 (1, 2). Bibliothèque numérique Caraïbe Amazonie Plateau des Guyanes: p. 80 (1). © Brenna Quinlan: p. 186 (1, 2). Chicago Historical Society: p. 62 (2). © Cliffor Harper: p. 152 (1) © Colin Moorcraft: p. 148 (3). © Constant Nieuwenhuys, Pictoright 2019: p. 70 (1, 2). Cornell University Library, Rare Book and Manuscript Collections: p. 118 (1, 2). Courtesy of David Holmgren: pp. 160 (2); 170 (1, 2), 156 (1, 2, 4); 174 (1), 190 (3), 192 (4). © Dogma: p. 79. © EcoLabs: p. 160 (1). Essam Al-Sudani: p. 46 (1). Fondo Antiguo de la Biblioteca de la Universidad de Sevilla: p. 110 (1). © Fondazione Musei Senesi: p. 54 (1, 2). © 2019, Frank Lloyd Wright Foundation, Scottsdale, AZ, all rights reserved: p. 87. © George Steinmetz: pp. 92 (3, 4), 94 (5). © Gruen Associates: p. 65. Harvard University, Houghton Library (modbm_soc_860_05):

p. 83. Courtesy of Howard Odum: p. 140 (2). Courtesy of Howard and Elisabeth Odum: pp. 140 (3), 142 (4, 5). © Italian Archeological Mission in Eastern Anatolia: p. 46 (3). © James C. Scott: p. 106 (1). © Jay Kinney: p. 147. © Jean-François Batellier: p. 91. © José Ortiz Echague, VEGAP 2019: p. 20 (1). © Julie Nahon: p. 133. © Kisho Kurokawa architect & associates: p. 70 (3, 4). © Le Corbusier, ADAGP 2019: pp. 23, 61, 66 (1), 159. Lisbon Architecture Triennale: p. 18 (4, 5). From the film Look & See: A Portrait of Wendell Berry: pp. 122 (2), 124 (3). © Lucas Wolf: p. 176 (2). © Martin Etienne: pp. 143, 196, 200, 204, 208. Courtesy of Mathieu Calame, © Le Basic (www.lebasic.com): p. 102 (5). © Michael Wolf: p. 117. © Ministère de la Culture – Médiathèque de l'architecture et du patrimoine, Dist. RMN-Grand Palais / François Kollar: p. 88 (1). © Murray Bookchin: p. 114 (4, 5). Muzeum umění Olomouc (inv. no. A7): p. 66 (2). © National Geographic: p. 92 (2). Courtesy of the New Alchemy Institute: p. 121. © OMA: pp. 52 (4), 155. © Pattern Language: p. 139. © Perrine & Charles Hervé-Gruyer: p. 182 (1, 2). © Portola Institute Inc.: pp. 126 (2), 128 (4). Prelinger Library: p. 114 (1, 2, 3). Pressephotos: p. 88 (2). Courtesy of Reda Erraziqi & Rose Hewins: pp. 169, 173, 177, 181, 185, 189, 193. Courtesy of Rob Hopkins: p. 190 (1, 2). © Roger B. Ulrich, Ostia Antica, 1985, Italia: p. 50 (1). © Sascha Richter: p. 108 (2, 3, 4). Scanned by Science et Société: p. 136 (6). Courtesy of Stewart Brand: p. 126 (3). Courtesy of SUB Göttingen: p. 31. © Superstudio: p. 99. © Stan Eales (www.staneales.com): p. 95. Courtesy of The Land Institute, Terry Evans: p. 130 (2). © The LIFE Picture Collection, Getty Images: p. 90 (6). © Thomas Jefferson Foundation at Monticello: p. 30 (3, 5). © Thoronet Cistercian Abbey, Centre des Monuments Nationaux: p. 110 (2). Courtesy of the Thünen Museum, Tellow. © Michael Welling: p. 60 (2). © Time Magazine: pp. 136 (5), 144 (3). © Ungers Archive for Architectural Research: p. 125. United States Department of the Interior, National Park Service, Frederick Law Olmsted National Historic Site: p. 68 (4). Courtesy of University Prints, Boston: p. 20 (3). Scanned by the University of Toronto: p. 78 (3). © US Naval Institute Photo Archive: p. 109. Courtesy of Wendell Berry: p. 122 (1). Courtesy of Wes Jackson: p. 130 (1, 3, 4). © Wiley: p. 148 (1, 2). Courtesy of Wladyslaw Sojka (www.sojka.photo): p. 69.

THE POETICS OF REASON

Curatorial Team
Éric Lapierre – *Chief Curator*, Ambra Fabi, Fosco Lucarelli, Giovanni Piovene, Laurent Esmilaire, Mariabruna Fabrizi, Sébastien Marot Tristan Chadney – *Curators*, Claudia Mion – *Curatorial Assistant*

Deputy Executive Director
Manuel Henriques
Management Assistance
Helena Soares
Cláudia Rocha
Production
Isabel Antunes – *Coordination*, Beatriz Caetano, Carla Cardoso Carolina Vicente, Inês Vidal, Marta Moreira, Ricardo Batista, Sofia Baptista, Tiago Pombal
Communication
Sara Battesti – *Coordination*, Ana Guedes, Cláudia Duarte, Inês Revés, Raquel Guerreiro, Ricardo Chambel, Susana Pomba
Educational Service
Filipa Tomaz – *Coordination*, Letícia Carmo, Joana Martins
Fundraising and Partnerships
Joana Salvado, Joanna Hecker
Graphic Design and Art Director
Marco Balesteros (Letra)
Website
Marco Balesteros (Letra), Sara Orsi

Board of Directors
José Mateus – *Chairman*, Nuno Sampaio – *Vice-Chairman*, José Manuel dos Santos – *Member*, Maria Dalila Rodrigues – *Member*, Miguel Varela Gomes – *Member*, Pedro Araújo e Sá – *Member*
Supervisory Board
José Miguel Alecrim Duarte – *Chairman*, Miguel Luís Cortês Pinto de Melo – *Vice-Chairman*, Ricardo Ferreira

Advisory Board

Leonor Cintra Gomes – *Chairman*, Álvaro Siza Vieira, Ana Tostões, António Mega Ferreira, António Mexia, António Pinto Ribeiro, Augusto Mateus, Bárbara Coutinho, Bernardo Futcher Pereira, Cláudia Taborda, Delfim Sardo, Diogo Burnay, Eduardo Souto de Moura, Fernanda Fragateiro, Filipa Oliveira, Gonçalo Byrne, Gonçalo M. Tavares, João Belo Rodeia, João Gomes da Silva, João Pinharanda, João Luís Carrilho da Graça, Jorge Figueira, Jorge Gaspar, Jorge Sampaio, José Monterroso Teixeira, José Fernando Gonçalves, José Manuel Pedreirinho, Luís Santiago Baptista, Luís Sáragga Leal, Manuel Mateus, Manuel Pinho, Maria Calado, Mark Deputter, Miguel Vieira Baptista, Miguel Von Hafe Pérez, Nuno Crespo, Nuno Grande, Pedro Baía, Pedro Bandeira, Pedro Gadanho, Raquel Henriques da Silva, Sérgio Mah

This book is part of the main exhibition entitled *Agriculture and Architecture: Taking the Country's Side* showcased at Garagem Sul in the context of *The Poetics of Reason*, as the fifth edition of the Lisbon Architecture Triennale (3 October to 2 December 2019).

The Lisbon Architecture Triennale is a non-profit association whose mission is to research, foster and promote architectural thinking and practice. Founded in 2007, it holds a major forum every three years for the debate, discussion and dissemination of architecture that crosses geographic and disciplinary boundaries.

Concept and Organisation

Trienal de Arquitectura de Lisboa

Strategic Partners

LISBOA CÂMARA MUNICIPAL · fundação edp · FUNDAÇÃO MILLENNIUM BCP

Structure Financed by

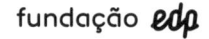

REPÚBLICA PORTUGUESA CULTURA · dg ARTES DIREÇÃO-GERAL DAS ARTES

Member

Future Architecture Platform|Member · Co-funded by the Creative Europe Programme of the European Union

Co-Producers

GARAGEM SUL EXPOSIÇÕES ARQUITECTURA · Culturgest Fundação Caixa Geral de Depósitos · FUNDAÇÃO CALOUSTE GULBENKIAN · maat · MUSEU NACIONAL DE ARTE CONTEMPORÂNEA DO CHIADO

Institutional Partner

Turismo de Lisboa

International Fund

Graham Foundation · AC/E ACCIÓN CULTURAL ESPAÑOLA

Partners

JRBOTAS · WMT · multiplacas

Associated Brands

panoramah! · Placo SAINT-GOBAIN · ESPORÃO · mpg · CIN

Media Partners

archdaily · Architectuul. · CANAL180.PT OUTRA TELEVISÃO · RTP · ANTENA 2

Associates

ABREU ADVOGADOS · babel · CASA DA ARQUITECTURA · CCB · fundação edp · FUNDAÇÃO MILLENNIUM BCP

/José Mateus arquitecto · ORDEM DOS ARQUITECTOS

High Patronage of His Excellency the President of the Portuguese Republic

O Presidente da República

Competition Support: Rock-Cultural Heritage Leading Urban Futures, Young Bird Plan.
Support: Cision; Dizplay; Space Collectors. **International Partners:** École d'Architecture Marne-la-Vallée, MIARD - Piet Zwart Institute, Fundació Catalunya-La Pedrera, Fondazione Aldo Rossi. **Hotel:** Hoteis Heritage Lisboa.
Triennale Patrons 2019/21: FSSMGN Arquitectos, Aurora Arquitectos, Colin Moorcraft.